Yifan He, W. Brent Lindquist, Svetlozar T. Rachev, and Davide Lauria
Risk Management for Cryptocurrency Portfolios

Yifan He, W. Brent Lindquist, Svetlozar T. Rachev, and Davide Lauria

Risk Management for Cryptocurrency Portfolios

DE GRUYTER

ISBN 978-1-5015-2009-9
e-ISBN [PDF] 978-1-5015-1713-6
e-ISBN [EPUB] 978-1-5015-1716-7

Library of Congress Control Number: 2025945958

Bibliographic information published by the Deutsche Nationalbibliothek
The Deutsche Nationalbibliothek lists this publication in the Deutsche Nationalbibliografie; detailed bibliographic data are available on the Internet at http://dnb.dnb.de.

Cover image: luza studios/iStock/Getty Images Plus

www.degruyterbrill.com
Questions about General Product Safety Regulation:
productsafety@degruyterbrill.com

Contents

Preface

Cryptocurrencies have emerged as a disruptive force in the financial world, offering new opportunities for investment and speculation. However, these opportunities come with unique risks that must be carefully managed. The volatile nature of cryptocurrency markets, coupled with regulatory uncertainties[1] and technological complexities, makes effective risk management essential for investors and traders in this space.

The motivation for writing this book arises from the growing need for a rigorous and systematic approach to risk management in cryptocurrency portfolios. While traditional risk management techniques, such as diversification and hedging, provide some level of protection, more sophisticated approaches relying on statistical analysis assuming non–normal, heavy–tailed distributions offer critical risk management enhancements. This book seeks to bridge that gap by offering a comprehensive quantitative approach tailored to digital assets.

Our primary goal is to provide readers with a deep understanding of risk management strategies specifically designed for cryptocurrency investments. By integrating advanced portfolio optimization techniques, heavy–tailed statistical modeling, and backtesting methodologies, we aim to equip practitioners and researchers with the necessary tools to navigate this evolving market with confidence.

This book is intended for a diverse audience, including financial professionals, quantitative analysts, academic researchers, and graduate students with an interest in mathematical finance and digital assets. While some familiarity with finance and statistics is beneficial, we have structured the content to be accessible to readers with varying levels of expertise. To facilitate practical application, we incorporate empirical case studies and MATLAB–based implementations in the book.

The structure of the book follows a logical progression to provide a comprehensive understanding of risk management and portfolio optimization in the context of cryptocurrency investments. Each chapter builds upon the previous one, gradually introducing more advanced concepts and techniques.

Chapter 1 introduces the dataset, consisting of 40 cryptocurrencies traded on the Binance exchange, used throughout the book. It provides a detailed analysis of the historical performance of these assets, highlighting their volatility and price trends over time.

Chapter 2 delves into the analysis of return time series, focusing on stationarity, the non-normal nature of cryptocurrency returns, and the heavy–tailed behavior of these returns.

---

1 At the time of completion of this book, the US Congress has just passed the GENIUS ACT, which outlines specific regulations under which banks and other entities can issue stablecoins. The intent of the act is to increase public trust in these assets and grow the crypto industry overall.

https://doi.org/10.1515/9781501517136-201

Chapter 3 explores modern portfolio theory, including mean–variance optimization, the capital market line, and conditional value–at–risk as a coherent risk measure.

Chapter 4 presents various portfolio optimization strategies, including long-only, long-short, and momentum strategies, along with performance evaluations using key metrics such as maximum drawdown and the Sharpe, Sortino–Satchell and Rachev ratios.

Chapter 5 extends the discussion to dynamic portfolio optimization, incorporating time–varying return distributions and copula–based models to more accurately capture the tail behavior of cryptocurrency returns.

Chapter 6 introduces robust optimization techniques, which treat the (multivariate) distribution of the observed (historical) random return vector as uncertain and subject to model misspecification.

Chapter 7 focuses on backtesting, introducing various validation methods, including value–at–risk tests, to assess the effectiveness of portfolio strategies discussed in previous chapters.

We would like to express our gratitude to Texas Tech University for support in the development of this book. Special thanks to De Gruyter for providing the platform to share our work with a wider audience. We hope this book serves as a valuable resource for those seeking to implement sound risk management practices in the cryptocurrency domain.

Lubbock, August 2025 — *Yifan He, W. Brent Lindquist, Svetlozar T. Rachev*
Bergamo, August 2025 — *Davide Lauria*

1 Cryptocurrency Dataset

This chapter focuses on the cryptocurrency dataset analyzed throughout this book, including the selection process, brief profiles of the selected assets, and an analysis of their historical performance.

1.1 Crypto Token Descriptions

The cryptocurrency dataset consisted of the 40 tokens traded on the Binance exchange having the highest market capitalization as of January 7, 2024. The choice of Binance reflects the fact that it is the cryptocurrency exchange with the highest trading volume. Founded in 2017, Binance is one of the largest cryptocurrency exchanges in the world.[1] It offers a wide range of digital currencies for trading and has grown rapidly due to its user-friendly platform, low trading fees, and high liquidity. Binance provides various services including spot trading, futures trading, staking,[2] and savings. It has its own cryptocurrency, Binance Coin (BNB), which can be used to pay for transaction fees on the platform at a discount. In 2020, Binance expanded its ecosystem with a second blockchain, Binance Smart Chain (BSC), upgraded to handle Web3 tools and smart contracts, allowing new decentralized services including blockchain games, governance and voting systems, and decentralized finance. Over the course of 2024, BSC replaced Binance's first generation BNB Beacon Chain.

Tab. 1.1 displays the ticker symbol, name, market capitalization, and year of earliest available analytic data for each cryptocurrency. A brief description[3] of each of the 40 cryptocurrencies is given below. The descriptions indicate the evolving range of uses for digital assets.

AAVE (Aave): Aave is a decentralized lending system built atop the Ethereum blockchain network. Founded in 2017, the Switzerland–based platform facilitates decentralized fi-

1 Throughout its history, Binance has been the subject of significant lawsuits and challenges from regulatory authorities. In some countries, notably the U.S. and U.K., the exchange has been banned from operating or ordered to cease operations. In order to comply with U.S. federal law, Binance has been replaced by a separate exchange, Binance, U.S., which however is also banned in six states. In November, 2023, Binance pleaded guilty to U.S. federal charges of money laundering, unlicensed money transmitting, and sanctions violations, resulting in a significant fine to the company, as well as a fine and prison term for, and replacement of, the Binance CEO.

2 For the beginner, think of staking as a way of committing some of your crypto coins to a pool run by validators, whose primary role is to participate in running the blockchain and maintaining its security. In return, as the validators add new blocks to the chain, you receive a percentage of each new block reward. Staking is only possible for cryptocurrencies linked to blockchains running a proof-of-stake consensus mechanism.

3 See https://www.binance.com/.

https://doi.org/10.1515/9781501517136-001

Tab. 1.1: Indicated data on the 40 selected cryptocurrencies as of Jan. 7, 2024.

Ticker	Name	Market Capitalization[a]	Volume[a] (24 h)	Analytics[b] Available
AAVE	Aave	1369.562	91.379	2020
ADA	Cardano	18406.403	584.579	2017
ALGO	Algorand	1351.107	43.996	2019
ATOM	Cosmos	3695.797	105.380	2019
AVAX	Avalanche	13342.839	762.158	2020
AXS	Axie Infinity	1027.187	69.450	2020
BAT	Basic Attention Token	338.245	14.274	2017
BCH	Bitcoin Cash	4732.836	169.341	2017
BNB	Build'N'Build	46212.497	820.521	2017
BTC	Bitcoin	853771.763	24655.637	2009
DOGE	Dogecoin	11622.747	389.524	2013
DOT	Polkadot	9249.694	336.396	2020
EGLD	MultiversX	1542.868	48.675	2020
ENJ	Enjin Coin	404.088	13.131	2017
EOS	EOS	794.012	98.220	2017
ETC	Ethereum Classic	3636.010	342.904	2016
ETH	Ethereum	278583.282	9572.661	2015
FIL	Filecoin	2630.008	140.198	2020
GRT	The Graph	1524.834	50.190	2020
HBAR	Hedera	2484.889	38.397	2019
ICP	Internet Computer	5466.879	132.724	2021
IOTA	MIOTA	799.090	25.479	2017
LINK	ChainLink	8736.920	482.856	2017
LTC	Litecoin	5054.875	264.651	2013
MATIC	Polygon	7709.958	354.479	2019
MKR	Maker	1812.335	50.637	2017
NEAR	NEAR Protocol	3084.291	219.402	2020
NEO	Neo	791.715	31.315	2016
QNT	Quant	1292.203	16.124	2018
SAND	The Sandbox	1032.995	83.602	2020
SHIB	Shiba Inu	5441.840	125.545	2021
SOL	Solana	46206.137	3522.991	2020
TRX	TRON	9870.746	244.515	2017
UNI	Uniswap	3701.221	80.417	2020
VET	VeChain	2113.248	41.247	2018
WAVES	Waves	252.925	32.617	2016
XLM	Stellar Lumens	3263.961	67.725	2014
XMR	Monero	2999.195	94.694	2014
XRP	Ripple	28594.836	1035.217	2013
XTZ	Tezos	964.185	38.556	2017

[a] In millions of BUSD [b] Source: CryptoRank.io

nance (DeFi) borrowing and lending of digital assets on the Ethereum, Polygon, and Avalanche blockchains. Its native crypto token AAVE can be traded on most exchanges or staked in the Aave platform to earn interest.

ADA (Cardano): Launched in 2017, Cardano is a decentralized proof–of–stake (PoS)[4] blockchain platform. It was conceptualized in 2015 by Charles Hoskinson, one of the co-founders of the Ethereum blockchain network.

ALGO (Algorand): Algorand is an open–source blockchain network and cryptocurrency designed to facilitate the mainstream adoption of blockchain technology. It aims to expand the use of digital currencies by accelerating the rates of transaction verification.

ATOM (Cosmos): Cosmos is a decentralized project and ecosystem developed to be an internet of blockchains. The vision for Cosmos is to provide developers with a fluid network that allows interoperability between blockchains and decentralized applications (DApps).

AVAX (Avalanche): Avalanche is a blockchain network that provides robust smart contract functionality and is designed to facilitate the use of DApps, non-fungible tokens (NFTs), and other complex blockchain platforms at scale.

AXS (Axie Infinity): Axie Infinity is a highly popular play–to–earn (P2E) game that allows players to earn blockchain–based tokens through gameplay. Axie Infinity Shards, or AXS, are the settlement token used within the Axie Infinity ecosystem to purchase in–game NFT assets.

BAT (Basic Attention Token): The Basic Attention Token was developed to enhance transparency in digital marketing, while rewarding network users, marketers and digital publishers in the process. BAT runs on the Brave decentralized web browser, which was built to track user attention in order to identify their interests and align them with relevant advertisers. Advertisers have the ability to set parameters to target the most relevant users according to matched interests, whom they can select for tiered rewards based on the accuracy of that targeting.

BCH (Bitcoin Cash): Bitcoin Cash is a cryptocurrency and blockchain network created as a result of a hard fork of the Bitcoin blockchain in 2017. Bitcoin Cash integrates a number of changes to the original Bitcoin code in order to provide faster transaction throughput and better scalability.

BNB (Build'N'Build): BNB is a cryptocurrency that can be used to trade and pay fees on the Binance exchange. It is the token that powers the BNB Chain ecosystem. It is one of the world's most popular utility tokens, enjoying a wide range of application.

---

4 The consensus mechanism used to achieve distributed agreement regarding the current state of the blockchain is a critical component of any cryptocurrency. While proof-of-work and proof-of-stake are, by far, the most prevalent, other consensus mechanisms exist. It is an interesting question as to whether this mechanism, which plays a fundamental role in the microstructure of a cryptocurrency, has any effect on its price dynamics. Addressing such a question is beyond the scope of this book.

BTC (Bitcoin): Bitcoin is one of the most popular cryptocurrencies in the market. First introduced in 2009, Bitcoin continues to be the top cryptocurrency according to market capitalization. It has paved the way for the many existing altcoins in the market and marked a pivotal moment for digital payment solutions.

DOGE (Dogecoin): Dogecoin is a novelty cryptocurrency originally launched as a "memecoin" within the cryptocurrency community. Over time, Dogecoin has grown into a large blockchain network and is one of the most popular altcoins available in the market.

DOT (Polkadot): Polkadot is a blockchain launched in 2016 by Gavin Wood, Ethereum's former CTO and co–founder. It allows developers to create customized interoperable parachains, or blockchains deployed from the Polkadot mainnet. Each parachain connects to the main relay chain, allowing them to communicate and share in the security measures of the mainnet.

EGLD (MultiversX): MultiversX is a blockchain–based protocol aiming to achieve comprehensive scalability by employing the full spectrum of sharding techniques[5] – network, transaction, and state. MultiversX positions itself as a versatile technological framework for the next generation Internet, encompassing DeFi, tangible assets, and the Metaverse domain. The platform for executing smart contracts is said to have the capacity for processing up to 100,000 transactions every second, complemented by minimal latency and comparatively lower transaction fees.

ENJ (Enjin Coin): Enjin is an Ethereum–based blockchain project designed to streamline and simplify access to NFTs and NFT markets. The Enjin ecosystem provides developers with the ability to create, distribute, and manage Ethereum–based NFTs.

EOS (EOS): EOS is the native token of the EOS blockchain. Launched in 2018, the EOS network provides a smart contract functionality that enables developers to build DApps. EOS is suited for NFT, game finance (GameFi), DeFi, and enterprise applications.

ETC (Ethereum Classic): Ethereum Classic is an open–source, decentralized computing platform and cryptocurrency launched in 2016 as an alternative to the Ethereum network. Ethereum Classic offers functionality for smart contracts and supports the deployment and use of DApps.

ETH (Ethereum): Ethereum is the second–largest cryptocurrency token in terms of market capitalization. This can be attributed to the innovation it has brought to the industry by introducing the functionality for smart contracts, which, in turn, has paved the way for DeFi and DApps.

5 Sharding is a technique the divides the blockchain network into smaller, independent partitions (shards). Each shard processes its own transactions, distributing the workload and increasing blockchain throughput.

FIL (Filecoin): Filecoin is a blockchain network and decentralized peer–to–peer (P2P) digital storage ecosystem and marketplace. Built on the InterPlanetary File System (IPFS) technology, Filecoin incentivizes sharing hard disk drives through its native FIL token.

GRT (The Graph): The Graph is an indexing protocol for querying data on networks such as Ethereum and IPFS, powering many applications in both DeFi and the broader Web3 ecosystem.

HBAR (Hedera): Hedera is an open–source, fast, PoS blockchain network and governance platform that allows developers to build and deploy DApps in the Solidity programming language. Hedera provides three services: a smart contract protocol; the hashgraph consensus algorithm, Consensus; and a token service for minting and deploying new cryptocurrencies and NFTs.

ICP (Internet Computer): Internet Computer operates as an internet for organizations and individuals to interact with one another. It enables them to use applications and software as they would on a centralized web platform such as Google or Yahoo, with one fundamental difference: it is decentralized.

IOTA (MIOTA): IOTA is an open–source distributed ledger and cryptocurrency designed to facilitate transactions between devices (sensors) and the global internet – the Internet of Things (IoT). It employs a directed acyclic graph (Tangle) to store ledger transactions rather than a blockchain based ledger. It does not use miners to validate transactions, nodes that issue a new transaction must validate two previous transactions. Consensus is achieved through a coordinator node operated by the IOTA Foundation, consequently the network is not completely decentralized.

LINK (ChainLink): Chainlink is an interoperability–focused blockchain network that aims to connect off–chain resources, applications, and data to blockchain ecosystems. Designed to facilitate interoperability between smart contract and real–world data, Chainlink makes it possible to authenticate and verify data for on–chain use.

LTC (Litecoin): Litecoin is a P2P payments cryptocurrency that was hard forked from Bitcoin in 2011. It was one of the first altcoins and allows users to send payments quickly and easily.

MATIC (Polygon): Polygon is a layer 2 scaling solution[6] built on the Ethereum blockchain, facilitating higher transaction throughput and lower transaction costs. Formerly known as the Matic Network, Polygon aims to create the blockchain infrastructure necessary to deploy and use Etheruem–based DApps at scale.

MKR (Maker): MKR is the utility token of the Maker platform, which functions as the backbone of the DAI stablecoin. As one of the first fungible (ERC–20) tokens to launch on the Ethereum network, DAI and Maker have a complex relationship governed through MKR tokens.

---

6 A layer 2 blockchain is a secondary network built on top of an existing blockchain (layer 1). Its primary purpose is to improve scalability and efficiency of the base layer blockchain.

NEAR (NEAR Protocol): NEAR Protocol is a blockchain network designed to facilitate the launch and operation of DApps. The NEAR Protocol ecosystem focuses on providing highly scalable functionality by integrating sharding as a layer 2 scaling solution.

NEO (Neo): Neo is a software ecosystem designed to function as a platform that can be used to create, deploy, and utilize DApps, services, and products. Frequently compared to Ethereum due to its robust smart contract functionality, Neo can be used to create DApps that support social media, prediction markets, DeFi and decentralized exchange (DEX) platforms.

QNT (Quant): Quant is a blockchain–based protocol designed to facilitate interoperability among different blockchain networks. It is primarily aimed at institutions such as banks and asset management firms, providing users with features that include blockchain–secured asset and data exchange, and cross–border payments.

SAND (The Sandbox): The Sandbox is a Web3, metaverse–based, decentralized game that users worldwide can play through any WiFi–compatible device. It is a P2E cryptocurrency gaming platform on which players are able to buy and sell metaverse land and NFTs. They can also develop, build, and sell their own unique virtual experiences within the gaming environment.

SHIB (Shiba Inu): Shiba Inu is a decentralized meme token, specifically a dog–themed cryptocurrency, created in August 2020 by an anonymous person or group of people under the pseudonym Ryoshi. It is based on the Ethereum blockchain.

SOL (Solana): Solana is an independent layer 1 blockchain created as a fast and efficient network with an underlying smart contract protocol. Since its launch in 2020, the Solana network has been marketed as a competitor to Ethereum.

TRX (TRON): TRON is a widely used public blockchains, claiming 100 million users and a cumulative total of over 3.4 billion transactions. The TRON network aims to decentralize content–sharing and establish a framework through which future Web3 platforms will operate.

UNI (Uniswap): Uniswap is a decentralized protocol built atop the Ethereum blockchain network. The protocol is an automated market maker (AMM), a system designed to facilitate the exchange of various ERC–20–based cryptocurrencies.

VET (VeChain): VeChain is a blockchain–based platform designed to decentralize the supply chain industry and function as a DApp ecosystem.

WAVES (Waves): Waves is a multi–purpose blockchain platform which supports various uses including DApps and smart contracts. Launched in June 2016 following one of the cryptocurrency industry's earliest initial coin offerings (ICO), Waves initially set out to improve on the first blockchain platforms by increasing speed, utility and user–friendliness.

XLM (Stellar Lumens): Stellar Lumens is a cryptocurrency used to facilitate cross–border digital payments on the blockchain. It is one of the oldest decentralized projects and cryptocurrency ecosystems, having been launched in 2014 by Ripple Labs co–founder Jed McCaleb.

XMR (Monero): Monero is a privacy–focused cryptocurrency that was launched in April 2014. It aims to provide secure, private, and untraceable transactions using advanced cryptographic techniques. Unlike Bitcoin, Monero utilizes ring signatures, stealth addresses, and confidential transactions to obfuscate transaction details such as the sender, receiver, and the amount transacted.

XRP (Ripple): Ripple is a decentralized payment network that was built to replace the traditional money transfer network, SWIFT. Ripple is a uniquely designed blockchain that works on a consensus model in which all nodes given access to the system must validate transactions on the network. This model ensures that payments on the network are subject to considerable security measures without compromised efficiency. Transactions on Ripple are typically completed within four to five seconds.

XTZ (Tezos): Tezos is a self–upgrading, open–source blockchain ecosystem designed to power Web3 services, protocols, and platforms such as DeFi applications, NFT marketplaces and projects, P2P payments, and smart contracts.

The cryptocurrency dataset consisted of
- exchange rates (prices) between each cryptocurrency and Binance US dollar (BUSD),[7]
- trading volume measured in BUSD, and
- the number of trades

within each one–minute period from 05:56:00[8] on July 29, 2021 to 05:05:00 on September 8, 2023,[9] amounting to 1,109,721 one–minute sample points[10] for each asset. These rates reflect the last realized transaction within each minute on Binance. As exchange rates may vary on other digital exchange platforms[11] (see, for instance, Makarov and Schoar 2020), we prefer to use actual transactions instead of averages across exchanges,

---

7 BUSD was a stable cryptocurrency designed to be at parity with the USD. It was issued by Paxos, a trust company based in New York City, and Binance. It was to be backed 1:1 by a reserve of U.S. dollars. As reported by Bloomberg in January 2023, BUSD "was often undercollateralized between 2020 and 2021. On three separate occasions, the gap between reserves and supply surpassed $1 billion." As a result, in February 2023 the New York Department of Financial Services issued an order to Paxos to stop minting new BUSD tokens. On September 1, 2023, Binance announced that it would discontinue support for BUSD, advising BUSD token holders to migrate to a new stablecoin, First Digital USD, with migration to be completed by February 2024. The timing of this discontinuation of BUSD does not affect our dataset.

8 The time zone corresponding to all timestamps in this dataset is UTC+0. We employ a 24 hour clock in reporting all times in this book.

9 The dataset has significant overlap with the period of the Covid–19 pandemic, which, for simplicity, we will consider as the period from January 2020 through December 2022.

10 Data is missing for three short periods: i) from 02:00:00 to 06:30:00 on August 13, 2021; ii) from 06:59:00 to 09:00:00 on September 29, 2021; and iii) from 12:39:00 to 14:00:00 on March 24, 2023.

11 Potentially allowing for arbitrage opportunities!

as our aim is to test trading strategies and risk management techniques, not to assess the general evolution of the values of the cryptocurrencies.

The time period covered by our data set (July 29, 2021 to Sept 8, 2023) contains significant cryptocurrency market events.[12]

T1) The years 2020–2021 experienced a cryptocurrency bubble. Prior to the beginning of our data set:

- In early 2021, BTC prices fluctuated wildly: rising above $34,700 on January 3, 2021 only to fall by 17% the next day; surpassing $50,000 on February 16, and $61,000 on March 13; plunging from $64,000 on April 14 to below $49,000 on April 23 (a 23% decline).
- On April 14, 2021 the crypto exchange Coinbase (San Francisco) went public; their share price grew by 31% on the first day creating a market capitalization for the exchange of $85.8B.
- By May 2021, the novelty token DOGE had increased in value by 20,000% in one year; shortly thereafter its value declined by 93%. By May 19, 2021, nearly all cryptocurrency values had declined by double–digit percentages (BTC by 30%, ETH by 40%). The exchange Binance went down for over an hour due to the volume of trades, preventing many traders from exiting positions.

T2) During the period of our data set:

- By early September, 2021 BTC again surpassed $52,000 while ETH grew by over 100% (surpassing $3,900).
- On October 19, 2021, the first US Bitcoin ETF "BITO" (ProShares) started trading. This fund debut was accompanied by heavy trading, attracting more than 1B USD in assets. However, by October 27, 2021, the ETF–related euphoria[13] had evaporated; traders cut crypto positions, and BTC fell by 5.3% on October 27.
- By November 7, BTC peaked at over $67,000, ETH at over $4,800.[14]

T3) After its November 2021 peak, the cryptocurrency market bubble collapsed, falling with the rest of the Covid era markets. Notably, during this period the exchange FTX (The Bahamas) collapsed and declared bankruptcy.

- By the end of 2021, BTC was down by 30% (to under $48,000) and ETH by 23% (to under $3,800).

12 Data in the itemized lists T1–T4 was summarized from the Wikipedia page on cryptocurrency bubbles, https://en.wikipedia.org/wiki/Cryptocurrency_bubble

13 Ossinger, Joanna and Hajric, Vildana. "Bitcoin slips below $60,000 as ETF–related bliss evaporates". Bloomberg, October 27, 2021.

14 For comparison with technology stocks, the NASDAQ peaked at just over 16,000 on November 19, 2021.

- On May 3, 2022, the 0.5% interest rate raise by the US FED triggered a market selloff. Over the next eight days BTC fell 27% (to just over $29,000), ETH by 33.5% (to just over $1,900).
- By May 10, the Coinbase share price was down 80% from its peak.
- May 5–13, the stable tokens Terra USD and Luna collapsed.
- June 12, the cryptocurrency company Celsius Network (New Jersey) indefinitely halted withdrawals and transfers. BTC fell 15% (to nearly $22,500), ETH fell to $1,200.
- June 13, the stable token USDD (Tron) collapsed.
- June 17, BTC dropped below $20,000, ETH below $1,000. The crypto lender Babel Finance (Hong Kong) froze withdrawals.
- June 23, the crypto exchange CoinFLEX (Seychelles) paused withdrawals.
- June 27, the cryptocurrency hedge fund Three Arrows Capital (Singapore), owing 3.5B USD, was forced into liquidation by a court in the British Virgin Islands. To protect its US assets, it declared bankruptcy in the US on July 2. Its US assets were frozen by the US Bankruptcy Court for the Southern District of New York on July 12.
- July 5, cryptocurrency broker Voyager Digital (New York) filed for bankruptcy.
- July 13, Celsius Network declared bankruptcy.
- July 19 and 20, the investment management firm SkyBridge Capital (New York) and crypto exchange Zipmex (Southeast Asia) froze withdrawals; the crypto exchange Vauld (Singapore) filed for bankruptcy.
- July 26, Coinbase shares fell 26% as the company came under SEC scrutiny.
- August 8, the cryptocurrency lender Hodlnaut (Singapore) suspended withdrawals.
- November 7 and 8, the FTX exchange experienced a withdrawal run; its token FTT lost 80% of its value.
- November 11, FTX declared bankruptcy.
- November 28, the digital asset lender BlockFi (New Jersey) declared bankruptcy.
- January 20, 2023, Genesis (parent company Digital Currency Group, Stamford Connecticut), a cryptocurrency intermediary for institutional investors, filed for bankruptcy.
- March 2–9, 2023, the shares of Silvergate Capital (San Diego), a banker to the cryptocurrency industry, plunged over 57%. Plans to liquidate its bank were announced.
- March 10–12, the New York State Department of Financial Services closed Signature Bank (New York City), another banker to the cryptocurrency industry, after its shares dropped 32%.
- August 15, the crypto exchange Dasset (New Zealand) went into liquidation.

T4) Following the end of the period of our data set:
- January 22, 2024, Terraform (creator of the Terra USD and Luna tokens) filed for bankruptcy.

1.2 Performance of Hourly Data

It is well known (see, e.g., Lo 1991) that intraday price data contains long–term memory dependence, whose effects increase as the time interval between price changes decreases. To reduce these effects, we transformed the original dataset into one–hour intervals as follows. Beginning at 06:00:00 on July 29, 2021, the data set was partitioned into non–overlapping, contiguous windows of 60 minutes.[15] The missing data blocks were aligned with the hourly windows as illustrated in the following example. The first missing block of data covered the time period from 02:00:00 to 06:30:00 on August 13, 2021. Consequently the available data from 06:30:00 to 07:00:00 was also dropped, leaving no data for the hourly windows with time stamps 03:00:00 to 07:00:00.[16] Total volume and number of trades were summed in each hour period. The price associated with each hour–window was the price recorded in the final minute of each hour. As a result, the transformed dataset has 18,496 timestamps. Unless otherwise specified, we use this one–hour dataset for the analyses in this book.

Figs. 1.1–1.3 display the hourly price evolution of these 40 cryptocurrency assets. For accurate comparison, we assumed an initial investment of 1 BUSD in each cryptocurrency. The times of notable market events from the list T2 and T3 are indicated on the plots. For comparison, Fig. 1.4 shows how the S&P 500 index (^SPX) (normalized to have unit value at the close of trading on July 28, 2021) fared over the same time period.

In contrast to the hourly crypto data, the ^SPX data in Fig. 1.4 is daily. We consider the behavior of ^SPX as reflective of the broad equities market. Taking into account the different y–axis scales of the panels in Figs. 1.1–1.3, the broad equity market held relatively constant compared to the crypto market; cumulatively outperforming over 3/4 of all crypto assets over this period. Of course, the equities market did not experience (the same level as) the cryptocurrency bubble at the end of 2021. ^SPX also experienced its peak value later (January 3, 2022) than the crypto tokens. Just as for crypto assets, the equity market experienced the May 3–11, 2022 sell–off due to the FED interest rate hike. It also suffered a decline during June 12–27, 2022 as weakness of cryptocurrency exchanges, tokens, and related companies were exposed. The equity market enjoyed an

15 Specifically, the first hour contained the 60 minute intervals having timestamps 06:01:00 through 07:00:00, the second window consisted of the timestamps 07:01:00 through 08:00:00, etc. The first hour was thus timestamped as 07:00:00 on July 29, 2021, the second as 08:00:00 on July 29, 2021, etc.

16 With the 02:00:00 minute data point missing, the hourly window comprised of data values with the time stamps 01:01:00 through 02:00:00 had only 59 values. Rather than delete this entire hour, we utilized the 59 values.

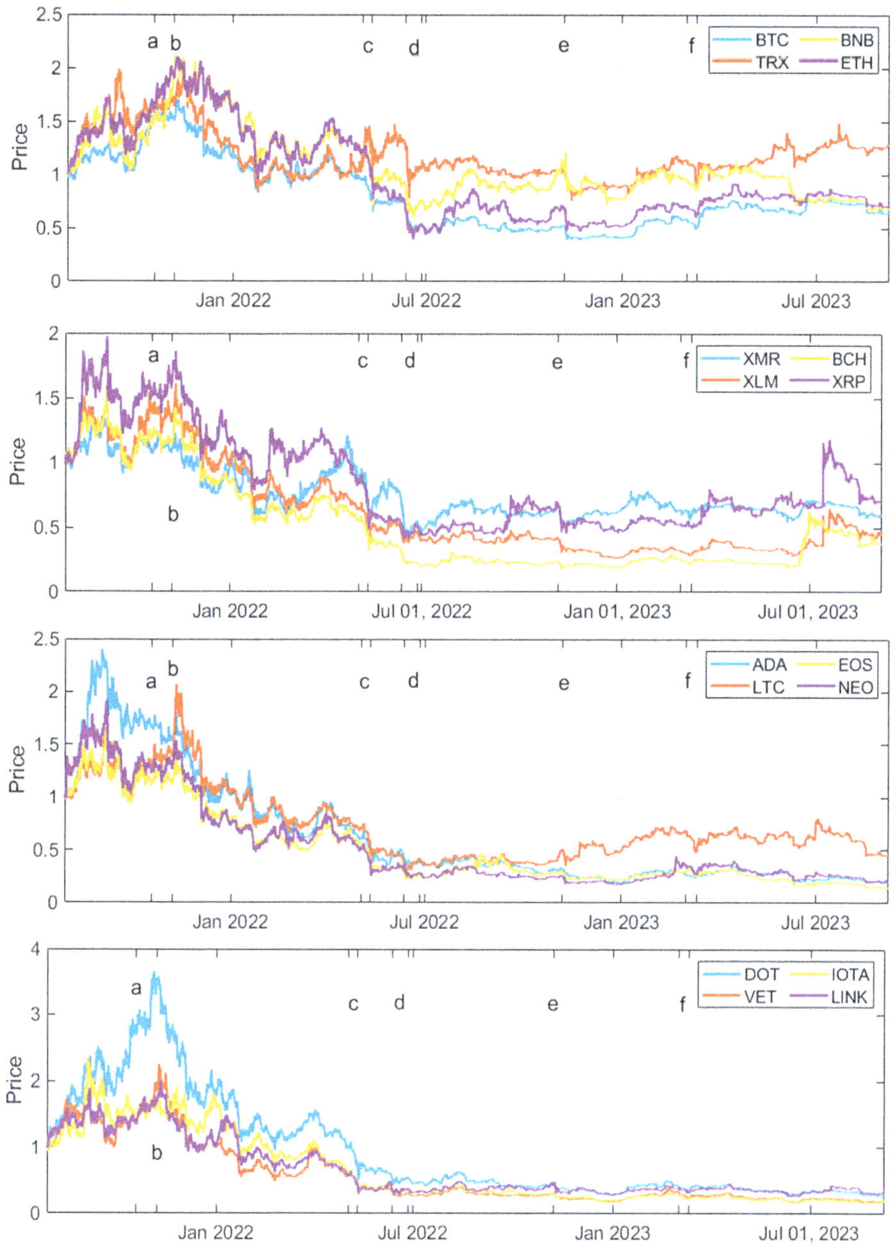

Fig. 1.1: Hourly prices of the indicated cryptocurrencies assuming an initial value of 1 BUSD on July 29, 2021 at 06:00:00 (a: Oct. 19 ETF; b: Nov. 7 peak; c: May 3–11 interest rate sell-off; d: June 12–27 collapses; e: Nov 7 FTX collapse; f: March 2–12 crypto bank closings)

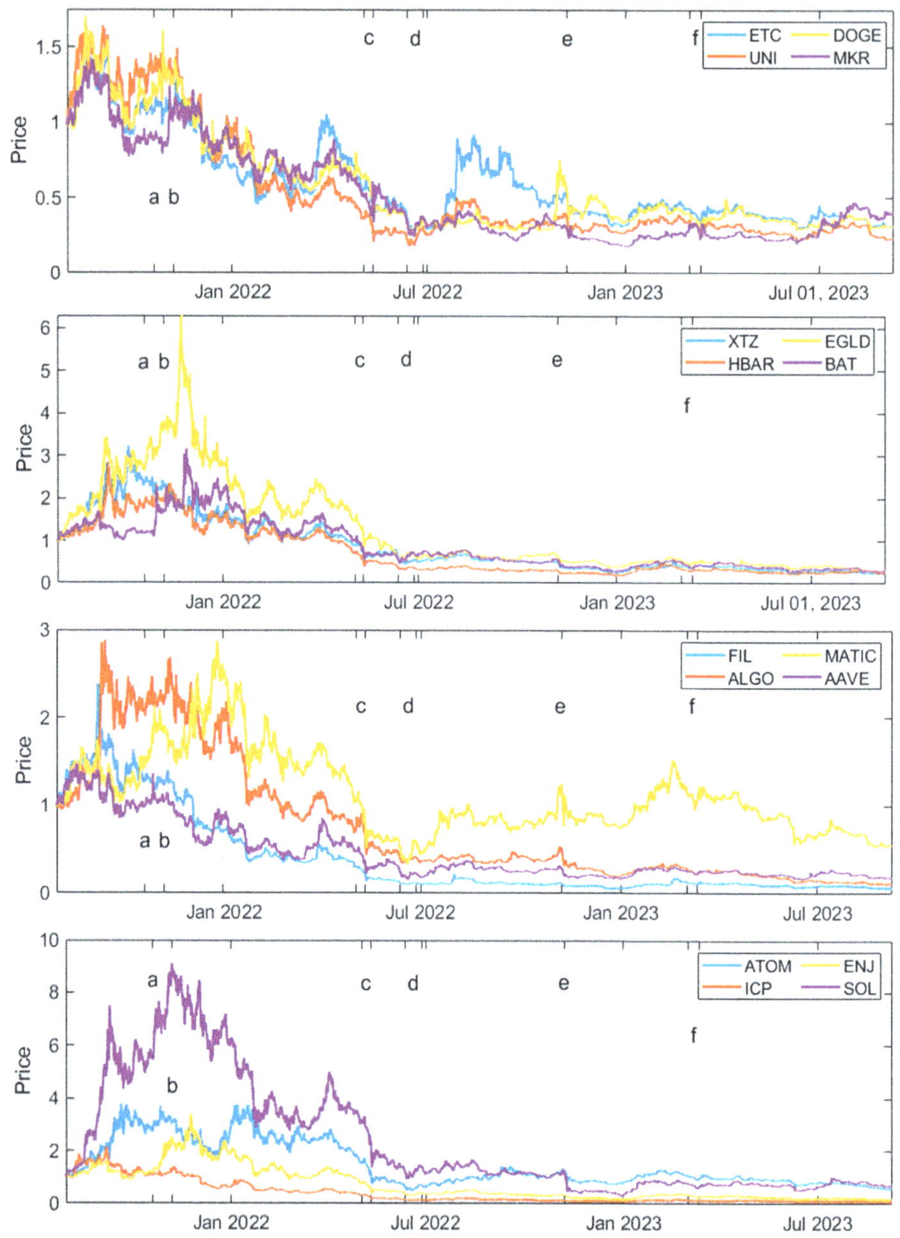

Fig. 1.2: Hourly prices of the indicated cryptocurrencies assuming an initial value of 1 BUSD on July 29, 2021 at 06:00:00 (a: Oct. 19 ETF; b: Nov. 7 peak; c: May 3–11 interest rate sell-off; d: June 12–27 collapses; e: Nov 7 FTX collapse; f: March 2–12 crypto bank closings)

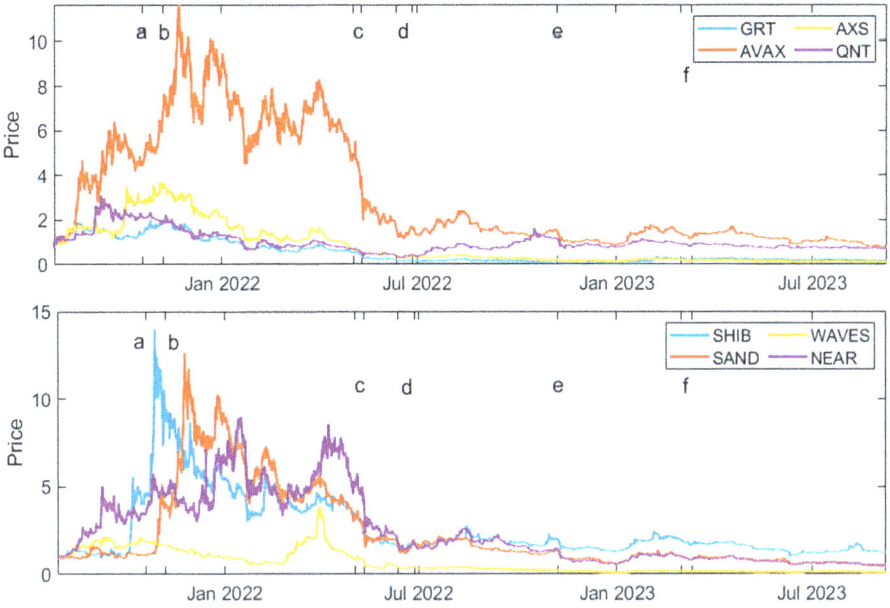

Fig. 1.3: Hourly prices of the indicated cryptocurrencies assuming an initial value of 1 BUSD on July 29, 2021 at 06:00:00. (a: Oct. 19 ETF; b: Nov. 7 peak; c: May 3–11 interest rate sell-off; d: June 12–27 collapses; e: Nov 7 FTX collapse; f: March 2–12 crypto bank closings)

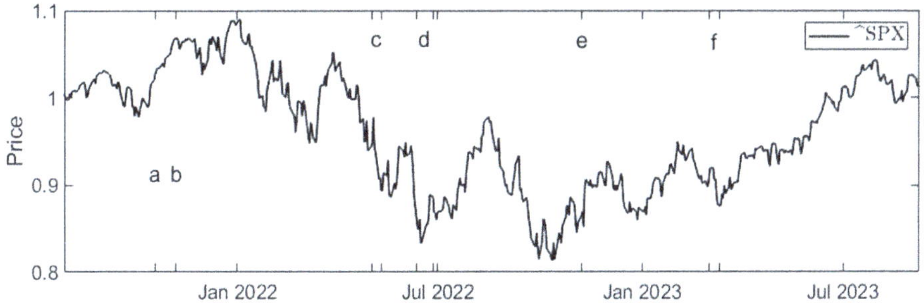

Fig. 1.4: Daily closing values of ^SPX assuming an initial value of 1 on July 29, 2021 at 06:00:00. (a: Oct. 19 ETF; b: Nov. 7 peak; c: May 3–11 interest rate sell-off; d: June 12–27 collapses; e: Nov 7 FTX collapse; f: March 2–12 crypto bank closings)

increase as a result of the FTX collapse on November 7, 2022. It suffered a small decline as the weaknesses of Silvergate Capital and Signature Bank operations in the crypto markets were exposed.

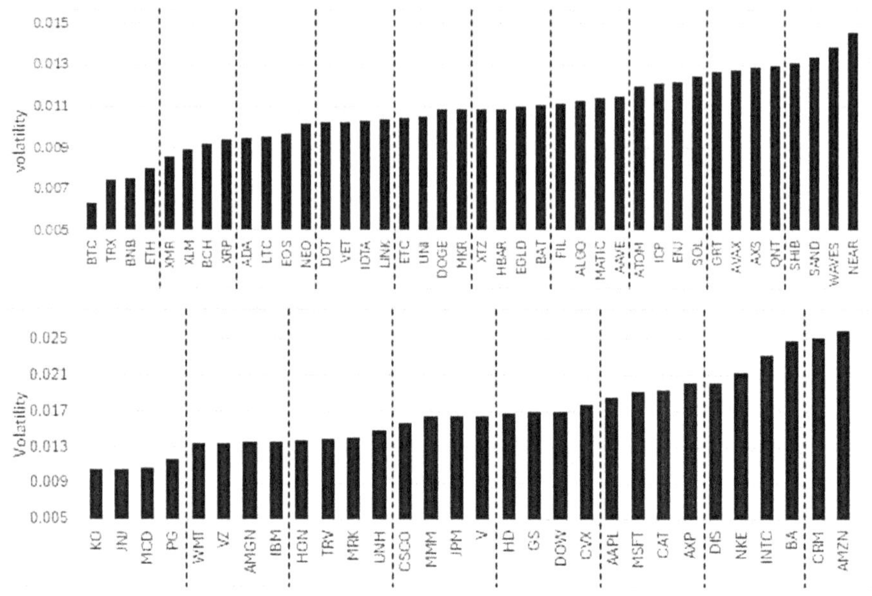

Fig. 1.5: Volatility of the (top) cryptocurrencies and (bottom) components of ^DJI

The crypto price time series in Figs. 1.1–1.3 were grouped into the panels in order of increasing volatility.[17] Fig. 1.5 displays the volatility of each cryptocurrency over the study time period. For comparison, the volatilities of the components of the Dow Jones Industrial Index (^DJI) over this time period are also shown. Note that the crypto volatilities measure standard deviation of hour returns while those for ^DJI measure standard deviation of daily returns.

17 The volatility is the standard deviation of the arithmetic return time series for each cryptocurrency. See Chapter 2 for a discussion of returns.

2 Return Time Series

This chapter provides an in–depth analysis of the time series of asset returns, focusing on stationarity, their non–normal distribution, and analysis of tail behavior.

Section 2.1 introduces the concept of stationarity of a stochastic process and discusses the methods used to transform nonstationary time series data into stationary data through differencing. The section explains the calculation of arithmetic returns and logarithmic returns (log–returns) and their use in financial models. It introduces the concept of the weight of an asset in a portfolio and covers the construction of an equal–weighted portfolio and the simpler equal–weighted buy–and–hold portfolio, whose prices serve as benchmarks against which to judge the performance of optimized portfolios. Finally, it introduces the concept of the distance between two time series and applies this to asset and benchmark return series.

Section 2.2 emphasizes the importance in financial analysis and risk management of understanding the distribution of the returns. It highlights the non–normal nature of cryptocurrency returns through the use of kernel density comparison, quantile–quantile plots, and statistical moments. The section also explores the heavy–tailed nature of the returns through the fitting of the generalized Pareto distribution and the Hill method, demonstrating the significant tail risk in cryptocurrency returns.

2.1 Stationarity

A stochastic process is stationary if its unconditional joint probability distribution and all its distribution measures do not change over time. A process is weakly stationary if the mean and covariance of the random variable do not change over time. Most time–series analyses (Manuca and Savit 1996; Kirchgässner et al. 2012; Hamilton 2020) rely on the assumption of weak stationarity. Using nonstationary time–series data in financial models can produce unreliable or spurious results, leading to poor forecasting (Chan 2006; Andersen et al. 2009; Ghouse et al. 2021). Nonstationary processes are of two types, either trend or difference stationary. Trend–stationary processes possess a deterministic, time–dependent mean value which can be removed. Stochastic shocks in such processes have only transient effects, after which the process reverts toward the deterministically evolving mean value. Difference stationary processes have characteristic equations possessing one (or more) unit roots. Stochastic shocks in such processes have permanent effects. Difference stationary processes with d unit roots must be differenced d times in order to make them stationary.

Visual examination of the cryptocurrency price series in Figs. 1.1 through 1.3 show no evidence of trend–stationarity. The series were therefore tested for the presence of

https://doi.org/10.1515/9781501517136-002

unit roots using the augmented Dickey–Fuller (ADF) test[1] (Dickey and Fuller 1979). The results of the ADF test are displayed in Tab. 2.1, with all values rounded to three decimal places. For each cryptocurrency, the p–value exceeds a significance level of 0.05 (equivalently, the test statistic value is greater than the critical value[2] of -1.9416) indicating that, at a 95% confidence level, we cannot reject the null hypothesis that difference stationarity exists in each time series.

Tab. 2.1: The results of the ADF test on each cryptocurrency price series

Ticker	Test Statistic	p-Value	Ticker	Test Statistic	p-Value
AAVE	-1.763	0.074	ADA	-1.218	0.206
ALGO	-1.241	0.198	ATOM	-1.003	0.285
AVAX	-0.978	0.294	AXS	-1.099	0.250
BAT	-1.186	0.218	BCH	-1.504	0.124
BNB	-0.718	0.389	BTC	-0.872	0.333
DOGE	-1.448	0.138	DOT	-1.012	0.281
EGLD	-1.035	0.273	ENJ	-1.328	0.171
EOS	-1.624	0.099	ETC	-1.423	0.145
ETH	-0.738	0.382	FIL	-1.622	0.092
GRT	-1.427	0.144	HBAR	-1.212	0.208
ICP	-1.735	0.079	IOTA	-1.288	0.183
LINK	-1.256	0.193	LTC	-1.145	0.233
MATIC	-0.978	0.294	MKR	-1.537	0.117
NEAR	-1.240	0.198	NEO	-1.464	0.134
QNT	-1.100	0.249	SAND	-1.150	0.231
SHIB	-1.289	0.183	SOL	-0.892	0.325
TRX	-0.310	0.538	UNI	-1.481	0.130
VET	-1.365	0.161	WAVES	-1.335	0.169
XLM	-1.246	0.196	XMR	-1.035	0.273
XRP	-0.894	0.325	XTZ	-1.161	0.227

As the ADF test gives no indication of the number of unit roots in a time series, we started with the assumption of a single unit root and applied first differences to the time series of the prices. Let $P_i(t)$ denote the price for asset i at time t (in this case measured in

1 The ADF test is available as the function `ur.df()` from the package `urca` in the software R (Berlinger et al. 2015). In MATLAB, it is available as the function `adftest()`. In Python, it is available as the function `adfuller()` from the package `statsmodels.tsa.stattools` (Kaabar 2024). Our computations were run using MATLAB.

2 For a given significance level, the critical value equals a constant value plus terms that are inversely proportional to powers of the sample size. Given the large sample size of $N = 18,496$ for each cryptocurrency, computation of the critical value for each yielded the same constant value of -1.9416 associated with the 0.05 significance level.

hours). First differences can be applied to the price or to the logarithm of the price. Each choice yields a "return" time series, either the arithmetic (discrete) return $R_i(t)$ or the log–return (continuous return) $r_i(t)$,

$$R_i(t) = \frac{P_i(t) - P_i(t-1)}{P_i(t-1)} = \frac{P_i(t)}{P_i(t-1)} - 1, \tag{2.1a}$$

$$r_i(t) = \ln(P_i(t)) - \ln(P_i(t-1)) = \ln\left(\frac{P_i(t)}{P_i(t-1)}\right). \tag{2.1b}$$

The arithmetic return is defined as a normalized first difference, the log–return is "naturally" normalized. The two returns are related[3]

$$R_i(t) + 1 = \frac{P_i(t)}{P_i(t-1)} = \exp\left[r_i(t)\right]. \tag{2.2}$$

The approximation $r_i(t) \approx R_i(t)$ holds quite well when asset returns are less than 1%. However, this approximation should always be avoided in empirical work as there is never a guarantee that returns will remain sufficiently small.

ADF tests were run on the cryptocurrency return time series, both arithmetic and logarithmic. In every case, the p–value returned from the test was ≤ 0.001, indicating a rejection of the ADF null hypothesis at the 0.05 significance level (i.e. the return time series were stationary).

At the end of time period $(t-1, t]$, the closing price $P_p(t)$ of a portfolio p consisting of n assets is

$$P_p(t) = \sum_{i=1}^{n} n_i(t-1) P_i(t), \tag{2.3}$$

where $P_i(t)$ is the closing price of asset i and $n_i(t-1)$ is the number of shares of asset i determined at time $t-1$ and held constant over the period $(t-1, t]$. Equation (2.3) can be expressed in the recursive form,

$$\begin{aligned} P_p(t) &= \left\{ \sum_{i=1}^{n} \left[\frac{n_i(t-1) P_i(t-1)}{P_p(t-1)} \right] \frac{P_i(t)}{P_i(t-1)} \right\} P_p(t-1) \\ &= \left\{ \sum_{i=1}^{n} w_i(t-1) \frac{P_i(t)}{P_i(t-1)} \right\} P_p(t-1), \end{aligned} \tag{2.4}$$

where $P_i(t-1)$ is the closing price of the asset at the end of period $(t-2, t-1]$, $P_p(t-1)$ is the respective closing price of the portfolio, and $w_i(t-1)$ is defined as the weight of asset i.

The notation $w_i(t-1)$ introduced in (2.4) is natural, as it involves values fixed at time $t-1$. However, it is a standard convention to rewrite $w_i(t-1)$ as $w_i(t)$ to denote the weight that is applied to asset i over the trading period $(t-1, t]$. It is crucial to recognize

3 Together, (2.1a), (2.1b) and (2.2) emphasize that the fundamental quantity is the ratio $P_i(t)/P_i(t-1)$, which provides the basis for both the arithmetic and log-return.

that, under this convention, the value of $w_i(t)$ is computed using prices and (possibly rebalanced) number of shares established after trading is finalized at time $t-1$. The weight $w_i(t)$ holds until trading is finalized at time t.

Adopting this notational convention, writing (2.4) in the symmetric form

$$\frac{P_p(t)}{P_p(t-1)} = \sum_{i=1}^{n} w_i(t) \frac{P_i(t)}{P_i(t-1)} , \tag{2.5}$$

leads naturally to the definition of the portfolio arithmetic return,

$$\begin{aligned} R_p(t) &= \frac{P_p(t)}{P_p(t-1)} - 1 = \left[\sum_{i=1}^{n} w_i(t) \frac{P_i(t)}{P_i(t-1)} \right] - 1 \\ &= \sum_{i=1}^{n} w_i(t) \frac{P_i(t)}{P_i(t-1)} - \sum_{i=1}^{n} w_i(t) \\ &= \sum_{i=1}^{n} w_i(t)) \left[\frac{P_i(t)}{P_i(t-1)} - 1 \right] \\ &= \sum_{i=1}^{n} w_i(t) R_i(t) , \end{aligned} \tag{2.6}$$

and log–return,

$$\begin{aligned} r_p(t) &= \ln \left[\frac{P_p(t)}{P_p(t-1)} \right] = \ln \left[\sum_{i=1}^{n} w_i(t) \frac{P_i(t)}{P_i(t-1)} \right] \\ &= \ln \left[\sum_{i=1}^{n} w_i(t) \left\{ R_i(t) + 1 \right\} \right] \\ &= \ln \left[\sum_{i=1}^{n} w_i(t) e^{r_i(t)} \right] . \end{aligned} \tag{2.7}$$

Notice that the final equality in (2.6) requires the identity $\sum_{i=1}^{n} w_i(t) = 1$, i.e. all of the wealth of the portfolio is invested in the n assets. Just as for any asset, the portfolio returns satisfy

$$R_p(t) + 1 = \frac{P_p(t)}{P_p(t-1)} = \exp[r_p(t)] . \tag{2.8}$$

Consequently, the price $P_p(t)$ can be obtained by knowing $P_p(0)$ and either return series,

$$\begin{aligned} P_p(t) &= P_p(0) \prod_{s=1}^{t} (1 + R_p(s)) \\ &= P_p(0) \exp\left(\sum_{s=1}^{t} r_p(s) \right) . \end{aligned} \tag{2.9}$$

Equation (2.9) identifies the cumulative arithmetic and log–returns of the portfolio,

$$
\begin{aligned}
R_p[0,t] &= \frac{P_p(t)}{P_p(0)} - 1 = \left[\prod_{s=1}^{t}(1+R_p(s))\right] - 1 \\
r_p[0,t] &= \ln\left[\frac{P_p(t)}{P_p(0)}\right] = \sum_{s=1}^{t} r_p(s)\,.
\end{aligned}
\tag{2.10}
$$

The arithmetic return $R_p(t)$ can be compactly expressed using vector notation,[4]

$$
R_p(t) = \boldsymbol{w}^\top(t)\boldsymbol{R}(t)\,. \tag{2.11}
$$

Due to the non–linear nature of the definition of log–returns, (2.7) does not have a similar compact vector form. However, if the arithmetic return $R_p(t)$ is sufficiently small, we have the approximation

$$
\begin{aligned}
r_p(t) &= \ln[1 + R_p(t)] \\
&\approx R_p(t) = \sum_{i=1}^{n} w_i(t) r_i(t) = \boldsymbol{w}^\top(t)\boldsymbol{r}(t),
\end{aligned}
\tag{2.12}
$$

Similarly, if $R_p(s)$, $s = 1, \ldots, t$ are sufficiently small, the cumulative arithmetic return in (2.10) can be approximated as

$$
R_p[0,t] \approx \sum_{s=1}^{t} R_p(s)\,. \tag{2.13}
$$

While approximations (2.12) and (2.13) are often used, their use can lead to significant error. For example, that the "small" errors, introduced each time increment, are aggregated by using approximation (2.12) in (2.9). Thus, in empirical computations employing either arithmetic or log–returns, exact formulas should always be used. This is particularly necessary for the high volatility cryptocurrency portfolios.

2.1.1 Equally Weighted Portfolio Benchmarks

It is common to use an equally weighted portfolio as a benchmark against which to compare the performance of an optimized portfolio. We consider two versions, the equal–weighted portfolio (EWP) and an equal–weighted buy–and–hold (EWBH) portfolio.

Let p be an optimized portfolio of n assets. The EWP consists of the same n assets with $w_i(t) = 1/n$ for all t. From (2.4), (2.6) and (2.7), the price, arithmetic and log–returns

4 By default, a vector $\boldsymbol{v}$ is assumed to be a column vector, while its transpose $\boldsymbol{v}^\top$ is a row vector. Thus, the column vector $\boldsymbol{v}$, having elements $v_1, v_2, \cdots, v_n$, can be written as $\boldsymbol{v} = (v_1, v_2, \cdots, v_n)^\top$, and the row vector $\boldsymbol{v}^\top$ can be written as $\boldsymbol{v}^\top = (v_1, v_2, \cdots, v_n)$.

of the EWP are given by

$$P_{\mathrm{EWP}}(t) = P_{\mathrm{EWP}}(t-1)\left\{\frac{1}{n}\sum_{i=1}^{n}\frac{P_i(t)}{P_i(t-1)}\right\},$$
$$R_{\mathrm{EWP}}(t) = \frac{1}{n}\sum_{i=1}^{n}R_i(t)\,, \qquad r_{\mathrm{EWP}}(t) = \ln\left[\sum_{i=1}^{n}e^{r_i(t)}\right] - \ln(n)\,, \tag{2.14}$$

In an EWP, the number of shares of each asset must change with the asset price to ensure that the weight $w_i(t)$ remains constant. In contrast, for the EWBH portfolio, the number of shares remains fixed $n_i(t) = n_i(0) = n_i$, $i = 1, \dots, n$. The initial allotment of shares is chosen such that the initial asset weights $w_i(0) = 1/n$ (using the convention in (2.4) for the time labeling of the weights) are equal. Consequently

$$n_i = \frac{W_0}{nP_i(0)}, \quad i = 1, \dots, n\,, \tag{2.15}$$

where $P_p(0) = W_0$ is the initial wealth allocated to the portfolio. The “buy–and–hold” strategy implies that no further actions are taken after the initial investment. From (2.15), (2.4) can be written

$$P_p(t) = \frac{W_0}{n}\sum_{i=1}^{n}\frac{P_i(t)}{P_i(0)} = \frac{W_0}{n}\widehat{P}_p(t)\,. \tag{2.16}$$

The arithmetic return of an EWBH portfolio can therefore be written

$$R_p(t) = \frac{\widehat{P}_p(t)}{\widehat{P}_p(t-1)} - 1\,, \tag{2.17}$$

and the weights (again using the convention of (2.4)) are

$$w_i(t-1) = \frac{W_0}{nP_p(t-1)}\frac{P_i(t)}{P_i(0)} = \frac{1}{\widehat{P}_p(t-1)}\frac{P_i(t)}{P_i(0)}. \tag{2.18}$$

Both EWP and EWBH are widely used in practice due to their simplicity. Maintaining an actual EWP can generate extensive transaction costs. However, it can be generated computationally as a theoretical benchmark, having the attraction that the portfolio wealth is equally divided among all assets at all times. In contrast, the EWBH portfolio generates no transaction costs and is a practical investment strategy in its own right. Its price tends to be dominated by the higher priced assets.

We consider an EWP benchmark consisting of the 40 cryptocurrencies in our data set. Figs. 2.1 to 2.3 are a replot of Figs. 1.1 to 1.3 with the price of the benchmark portfolio added to each panel. It is tempting to consider the performance of individual cryptocurrencies relative to the benchmark portfolio over two distinct periods, prior to July 2022, and afterwards. Prior to July 2022, the performance of the EWP exceeds, or is competitive with, the performance of the 32 cryptocurrencies in Figs. 2.1 and 2.2, while it under performs seven of the eight cryptocurrencies in Fig. 2.3. However, continuing to hold each investment until September 2023, results in approximately 14 cryptocurrencies equaling, or outperforming, the EWP. It is therefore of interest to see how the price performance of optimized cryptocurrency portfolios would compare.

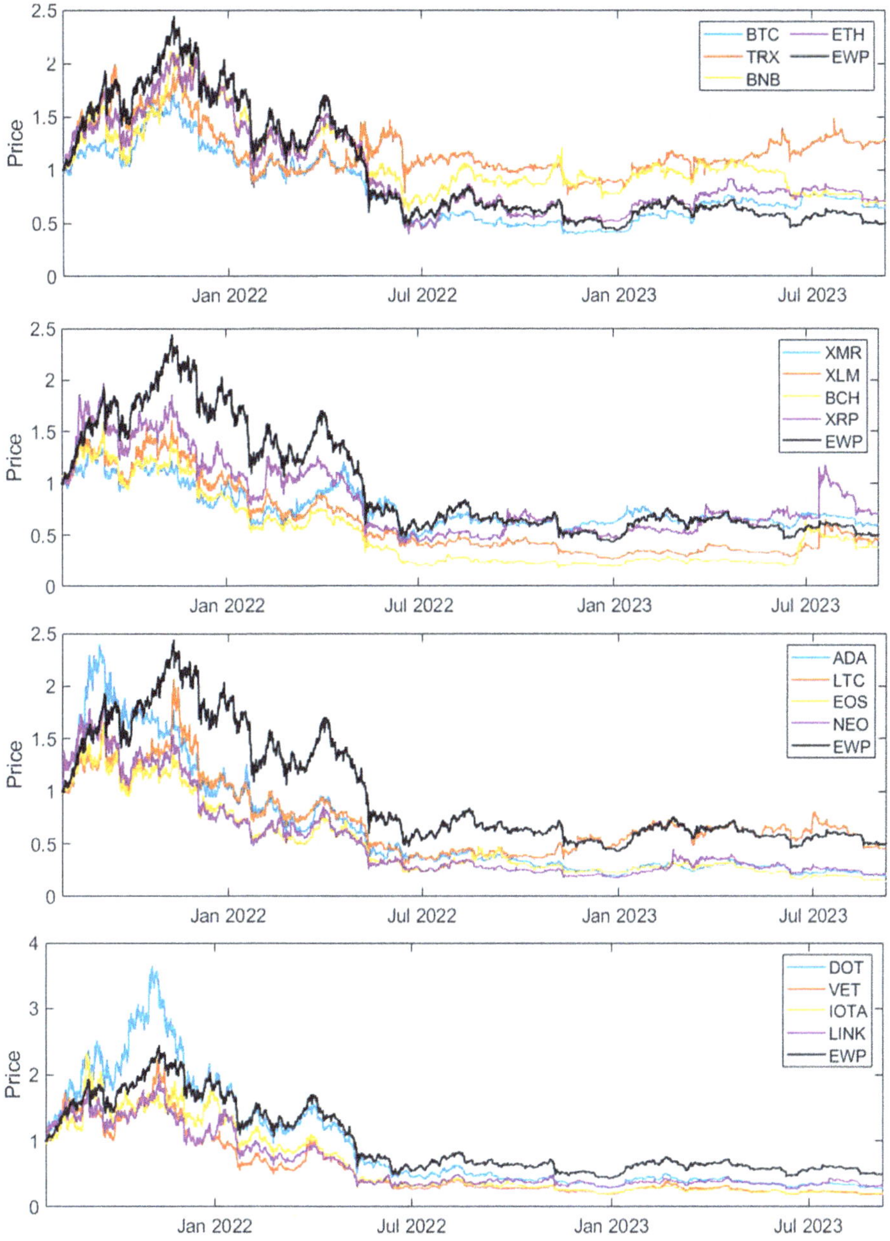

Fig. 2.1: Comparison of the hourly prices of the cryptocurrencies in Fig. 1.1 relative to the benchmark EWP

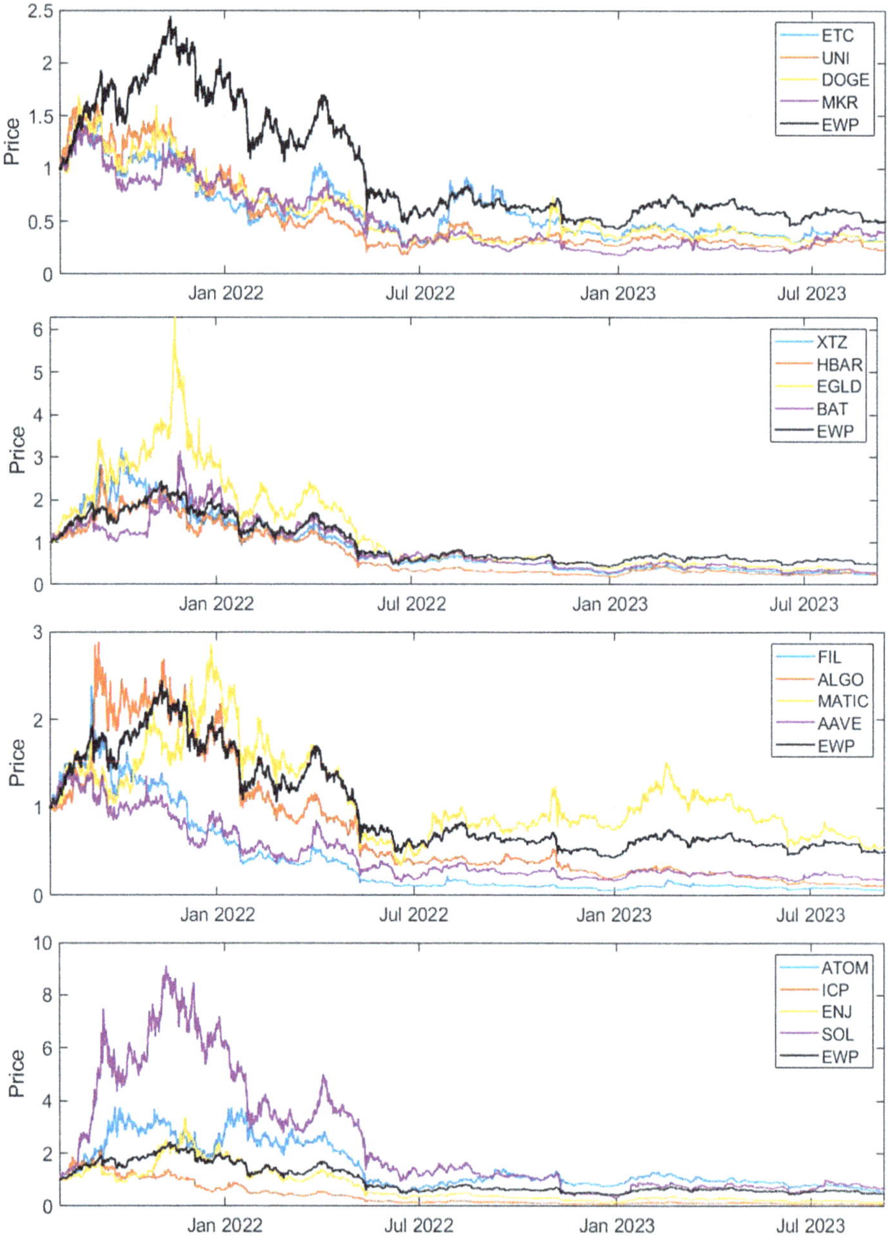

Fig. 2.2: Comparison of the hourly prices of the cryptocurrencies in Fig. 1.2 relative to the benchmark EWP

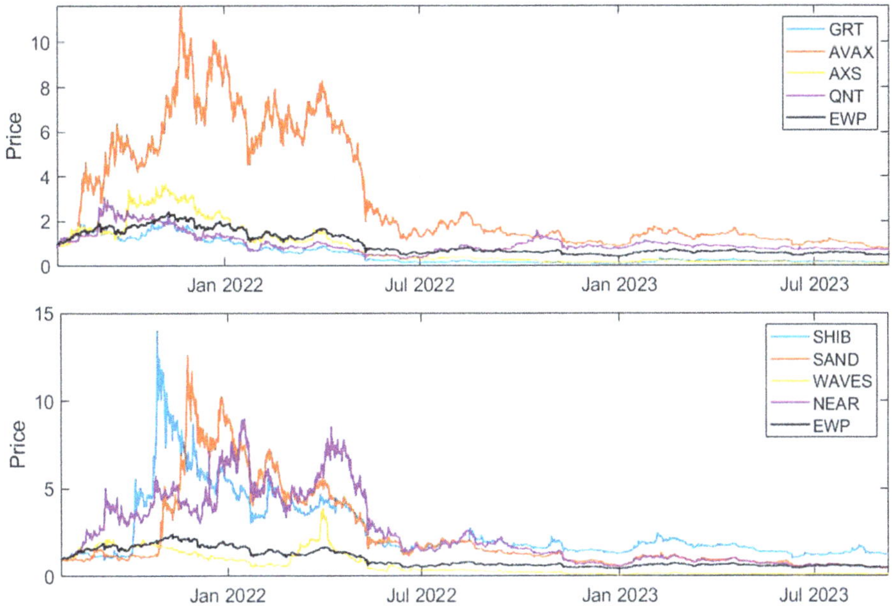

Fig. 2.3: Comparison of the hourly prices of the cryptocurrencies in Fig. 1.3 relative to the benchmark EWP

2.1.2 Distance between Returns

Let $X(t)$ and $B(t)$ denote two time series, where $B(t)$ is considered as a benchmark for $X(t)$. A distance time series between X and B can be defined as the difference[5]

$$d(X(t), B(t)) = X(t) - B(t) \ . \tag{2.19}$$

We apply (2.19) to return time series, specifically computing the distance series

$$\begin{aligned} d(R_i(t), R_{\mathrm{EWP}}(t)) &= R_i(t) - R_{\mathrm{EWP}}(t), \\ d(R_i(t), r_i(t)) &= R_i(t) - r_i(t), \end{aligned} \tag{2.20}$$

for cryptocurrency i. Comparisons of the form $d(R_i(t), R_{\mathrm{EWP}}(t))$ are a measure of the return difference between individual assets and that of the EWP benchmark. Comparisons of the form $d(R_i(t), r_i(t))$ measure the difference between employing arithmetic and log–returns, of particular concern if any computation approximates (2.2) by $1 + R(t) = e^{r(t)} \approx 1 + r(t)$. As $x - 1 - \ln(x) \geq 0$ for all $x > 0$, with equality holding only for $x = 1$, then $d(R_i(t), r_i(t)) \geq 0$. (This follows from (2.1a) and (2.1b) with $x = P_i(t)/P_i(t-1)$.)

5 We avoid the relative distance $d(X(t), B(t)) = (X(t) - B(t))/|B(t)|$, as this is numerically unstable when $|B(t)| << |X(t) - B(t)|$. We also avoid the more stable form $d(X(t), B(t)) = 2(X(t) - B(t))/(|X(t)| + |B(t)|$ as this distorts comparisons between $d(X(t), B(t))$ and $d(Y(t), B(t))$, where $X(t)$ and $Y(t)$ are different time series to be compared to $B(t)$. We also avoid the positive measure $d(X(t), B(t)) = |(X(t) - B(t))|$, as we are interested in the time-dependent sign of the difference.

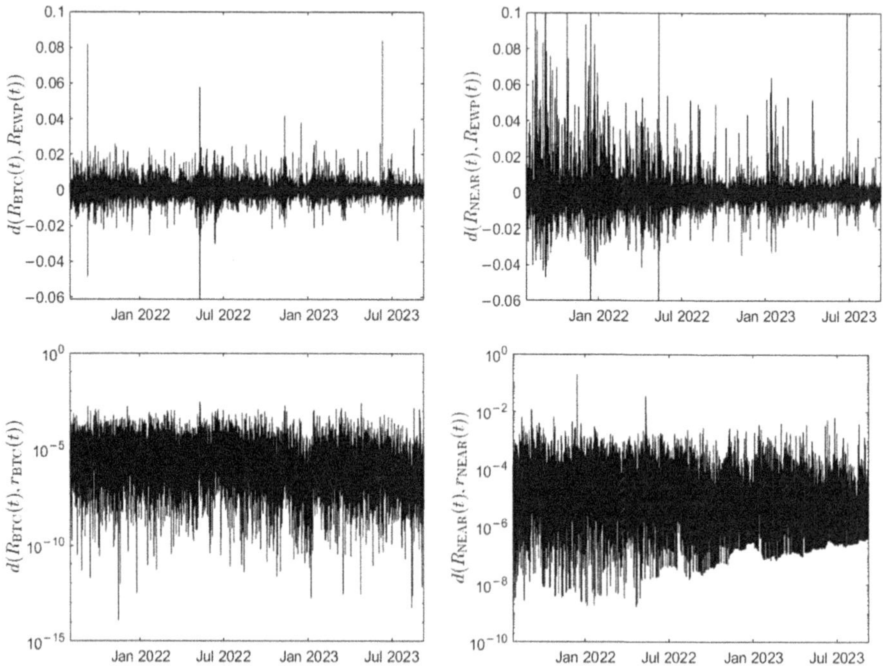

Fig. 2.4: Time series of select return distances. Note the y–axis log scale for the $d(R_i(t), r_i(t))$ plots

Fig. 2.4 displays the distance time series

$$\begin{aligned} &d(R_{\text{BTC}}(t), R_{\text{EWP}}(t)), \qquad &d(R_{\text{NEAR}}(t), R_{\text{EWP}}(t)),\\ &d(R_{\text{BTC}}(t), r_{\text{BTC}}(t)), \qquad &d(R_{\text{NEAR}}(t), r_{\text{NEAR}}(t)). \end{aligned}$$

The cryptocurrencies BTC and NEAR were chosen as they have, respectively, the smallest and largest volatility over the time period of the dataset, and thus provide a representative “bracket” for the distances for all of the cryptocurrencies in the dataset. It is clear that distances can change dramatically from hour to hour.

Fig. 2.5 displays box–whisker summary plots of the distributions of these four time series. The box–whisker summaries provide a more quantitative overview of the distribution of distances displayed in Fig. 2.4. Tab. 2.2 provides the numerical values for the whiskers and quartiles of each box–whisker summary.

More than 50% of the returns from these two cryptocurrencies are smaller than those for EWP; the median distance from these two cryptocurrencies to the benchmark EWP is negative, with the highest volatility cryptocurrency, NEAR, having the greater median distance. Reflecting its higher volatility, the inter quartile region and lower and upper whiskers are, respectively wider, than for the lowest volatility crypto asset, BTC. Note that, for visual clarity the y–axis on the two $d(\cdot, R_{\text{EWP}}(t))$ box–whisker plots has been clipped, eliminating some extreme outliers from view. Comparing the arithmetic

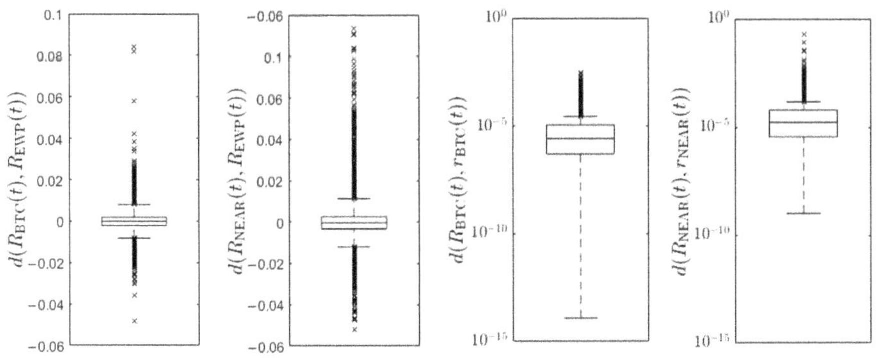

Fig. 2.5: Box–whisker summaries of the distribution of return distances in Fig. 2.4. Note the log scale for the $d(R_i(t), r_i(t))$ plots

Tab. 2.2: Quantile and whisker values for the distributions in Fig. 2.5

Distance	Scale[1]	Lower Whisker	Q_1	Q_2	Q_3	Upper Whisker
$d(R_{BTC}(t), R_{EWP}(t))$	10^{-3}	−8.19	−2.15	−0.159	1.88	7.91
$d(R_{NEAR}(t), R_{EWP}(t))$	10^{-3}	−12.0	−3.30	−0.459	2.53	11.3
$d(R_{BTC}(t), r_{BTC}(t))$	10^{-6}	$1.19 \cdot 10^{-14}$	0.504	2.74	11.6	28.2
$d(R_{NEAR}(t), r_{NEAR}(t))$	10^{-5}	$9.96 \cdot 10^{-10}$	0.360	1.70	6.17	14.9

[1] Each value provides the scale for all values on the row, with the exception of two lower whisker values, where the scale is provided explicitly.

and log–return time series for each cryptocurrency provides insight into how often the approximation $r_i(t) \approx R_i(t)$ might be justified (at a single time value). The distances between arithmetic and log–returns are roughly 10 times smaller (median value of order 10^{-6}) for BTC than for NEAR (median value of order 10^{-5}). Note however, that outliers can extend by factors of 100 to 1000 beyond the upper whisker.

The usual models for portfolio optimization (discussed in Chapter 3) are designed for discrete time steps. While the beautiful option pricing theory initiated by Black and Scholes (1973) and Merton (1973) relies on continuum stochastic partial differential equations to produce a continuous–time analytic form for option pricing under restricted assumptions, practical option pricing is performed using discrete tree models (Cox et al. 1979; Jarrow and Rudd 1983). For these reasons, we recommend that arithmetic returns, and their related formulas in (2.6), (2.9), (2.10), and (2.11) be used in empirical computations for portfolio optimization and option pricing.[6] However, we recognize that the

6 It is important to note that using arithmetic returns or log–returns in discrete option pricing models leads to different continuous–time limit processes under the Donsker–Prokhorov invariance principle (Hu et al. 2024; Lauria et al. 2023).

debate over whether to use arithmetic or log–returns is on–going. We refer the interested reader to Rachev et al. (2024, footnote 11) for a more detailed discussion, including references, on this subject.

In the text above, we have been careful to maintain a distinguishing notation for arithmetic and log–returns. For the remainder of this book, we drop that distinction and use $r_i(t)$ (or the vector notation $\boldsymbol{r}(t)$) to refer to return. Unless otherwise noted, the return referred to is always the arithmetic return.

2.2 Tail Behavior of Returns

Understanding the distribution of asset returns informs a wide range of financial decisions and risk management strategies.

- In risk management, it helps identify the risk of extreme losses (tail risk), enabling better portfolio construction and risk mitigation.
- In option pricing, the accurate modeling of the return distributions, taking into account the skewness and kurtosis observed in real markets, leads to improved pricing models (Hull 2022).
- For portfolio optimization, techniques like Markowitz's mean–variance optimization rely on understanding the return distributions to balance risk and return effectively, incorporating metrics such as value–at–risk and conditional value–at–risk (also known as expected shortfall) (Fabozzi and Markowitz 2011).
- Returns are deeply embedded in performance evaluation through reward–to–risk metrics such as the Sharpe ratio (Sharpe 1966).
- Return distributions are a critical component of stress testing, whereby financial institutions evaluate portfolio performance under extreme market conditions, ensuring better resilience and regulatory compliance (Jorion 2007).

The distribution of returns of financial assets often deviates significantly from the normal (Gaussian) distribution, exhibiting heavy tails and skewness. This phenomenon is particularly pronounced in the returns of stocks, commodities, and cryptocurrencies. The heavy–tailed nature of these return distributions implies a higher probability of extreme outcomes than what would be expected under a normal distribution. This poses significant challenges for risk management and financial modeling.

The heavy tails in asset returns can be attributed to various market dynamics, including sudden economic events, fluctuations in the market's liquidity, and investor behavior. In their seminal work on fractal market analysis, Mandelbrot and Hudson (2004) highlighted the fact that financial markets are prone to abrupt changes, often resulting in large price swings. These findings were supported by Cont (2001), who demonstrated that asset returns exhibit the properties of Lévy processes, characterized by jumps and heavy tails.

The heavy–tailed nature of asset returns has profound implications for risk management. Traditional models, such as the normal distribution based value–at–risk, often underestimate the risk of extreme events (tail risk). To address this, alternative risk measures like conditional value–at–risk have been proposed, which provides a more comprehensive view of tail risk by focusing on the average loss in the worst–case scenario (Rockafellar and Uryasev 2000).

In this section, through the use of graphical methods and hypothesis tests, we show that the distribution of returns is non–normal and heavy–tailed. We then use the generalized Pareto distribution and the Hill estimator to characterize that tail behavior.

2.2.1 The Non–Normal Nature of Returns

The question of whether a given distribution is normal[7] can be addressed by:

- comparing fits to the distribution of a non–parametric kernel density distribution against that of the normal distribution;
- quantile–quantile (QQ) plots; or
- measurement of the skewness and kurtosis (third and fourth central moments) of the distribution.

A QQ plot compares the relationship between the empirical quantiles of the sample distribution with the quantiles of a theoretical distribution.[8] The coordinates of the axis are chosen so that a linear relationship between the quantiles indicates that the empirical and theoretical distributions are the same. If the theoretical distribution is chosen to be the normal distribution, the plot is referred to as a normal QQ plot. For a normal QQ plot, the x–axis (normal distribution axis) is measured in terms of standard deviations.

Fig. 2.6 provides comparison of the return histogram with the kernel density and normal distributions for the cryptocurrencies having the four lowest and highest volatility. The kernel density distribution is designed to adapt to the empirical histogram; for our large data sets it provides very accurate fits. The normal distribution fits, however, are poor. The results shown are representative of all 40 cryptocurrencies.

7 The normal distribution is often attributed to the mathematician Carl Gauss who, in 1809, developed the formula for the distribution and showed that measurement errors were fit well by this distribution. Thus, the normal distribution is also referred to as the Gaussian distribution. However, the normal distribution formula was developed independently by the mathematician Robert Adrian in 1808, while the computational work of the 18'th century statistician Abraham de Moivre led to his "discovery" of a symmetric distribution as the large–n limit of the binomial distribution – the distribution that we now refer to as the normal distribution.

8 In contrast to the distribution density comparisons of Fig. 2.6, a QQ plot effectively compares the empirical and theoretical cumulative distributions.

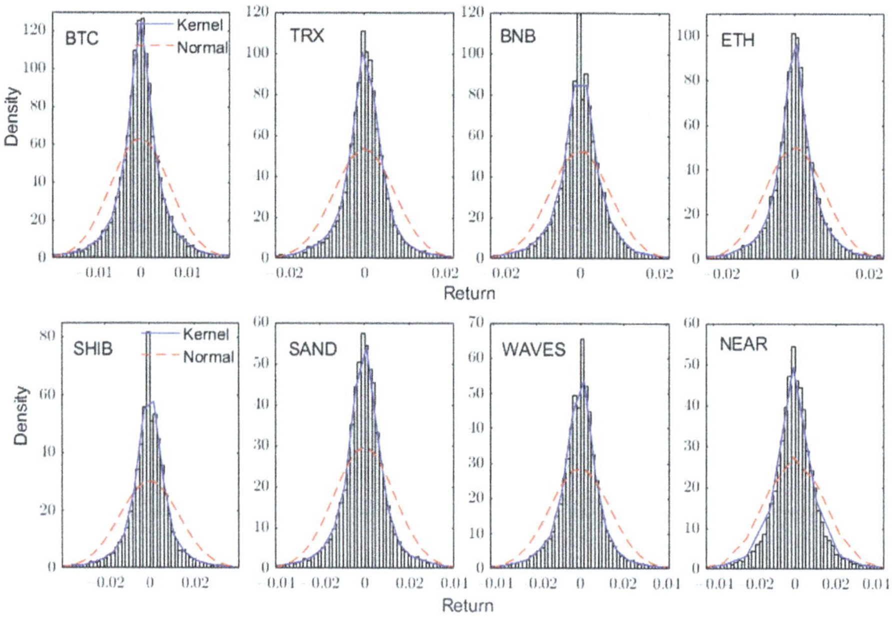

Fig. 2.6: Comparison of the return distribution histogram and the kernel density and normal distribution fits for the cryptocurrencies having, respectively, the four lowest and highest volatilities

Normal QQ plots for the cryptocurrencies analyzed in Fig. 2.6 are shown in Fig. 2.7. The tails of the return distributions (i.e., beyond $\sim \pm 1.5$ standard deviation) are decidedly not normal, exhibiting what are referred to as heavy tails.[9]

Fig. 2.8 presents the kernel density and normal distribution fit comparisons as well as the normal QQ plot for the EWP return distribution. As the individual cryptocurrencies have heavy–tailed distributions, an equally weighted sum of such distributions must also be heavy–tailed.

A normal distribution has skewness $\gamma = 0$ and kurtosis $\kappa = 3$. The difference $\kappa - 3$ is referred to as the excess kurtosis. Hypotheses tests are available to test deviations from normality due to the skewness, due to the excess kurtosis, or (in a combined test) to either the skewness or kurtosis (D'Agostino et al. 1990). We employed the `skewtest()`, `kurtosistest()`, and `normaltest()` functions from the Python subpackage `scipy.stats` to run these respective hypotheses tests. In each case, the null hypothesis is that the indicated moment(s) are consistent with the normal distribution.

The results for the individual skewness and kurtosis tests are presented in Tab. 2.3. The skewness test null hypothesis of normality was rejected at the 0.001 significance level in all cases except three, for the cryptocurrencies ADA, BTC, and ETH. At the 0.01

9 Heavy–tailed distributions exhibit large negative and positive values with greater probability than a normal distribution.

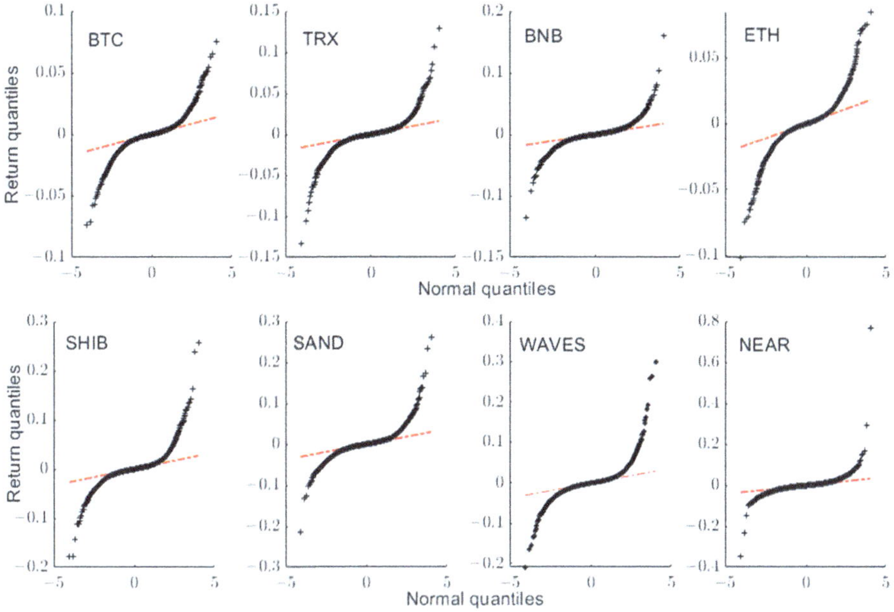

Fig. 2.7: Normal QQ plots of the return distributions for the selected cryptocurrencies

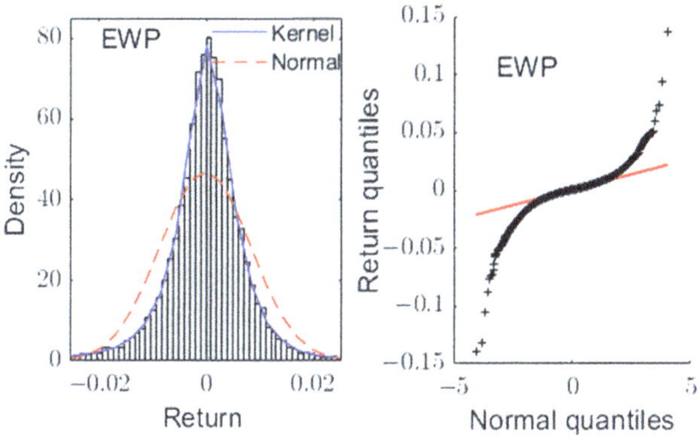

Fig. 2.8: (left) Kernel density and normal distribution fits and (right) normal QQ plot of the return distributions for the EWP

significance level, the null hypothesis was also rejected for BTC. For the excess kurtosis test, the null hypothesis of normality was rejected at the 0.001 significance level in all cases. Consequently, the null hypothesis of normality was rejected at the 0.001 significance level for all cases in the combined skewness–kurtosis test.

Summarizing the distribution fit comparisons, the QQ plots and the hypothesis test on the central moments, we have strong empirical evidence for rejecting a normal dis-

Tab. 2.3: The results of the test for normality on the skewness γ and excess kurtosis $\kappa - 3$ for each cryptocurrency and EWP

Ticker	γ (p-value)	$\kappa - 3^b$	Ticker	γ (p-value)	$\kappa - 3^b$
AAVE	0.269 (***)[a]	13.8	ADA	-0.006 (0.731)	14.6
ALGO	0.435 (***)	26.4	ATOM	0.302 (***)	23.8
AVAX	0.391 (***)	27.7	AXS	2.268 (***)	66.6
BAT	0.167 (***)	12.1	BCH	0.720 (***)	29.6
BNB	0.154 (***)	30.2	BTC	0.054 (0.003)	14.2
DOGE	1.326 (***)	41.2	DOT	0.090 (***)	25.6
EGLD	-0.153 (***)	18.5	ENJ	0.389 (***)	16.4
EOS	-0.710 (***)	25.8	ETC	-0.094 (***)	15.4
ETH	-0.028 (0.116)	13.4	FIL	-1.012 (***)	30.6
GRT	-0.192 (***)	11.0	HBAR	0.201 (***)	14.1
ICP	-0.066 (***)	13.7	IOTA	-0.555 (***)	15.1
LINK	-0.262 (***)	11.8	LTC	0.434 (***)	42.8
MATIC	0.200 (***)	13.2	MKR	1.096 (***)	28.0
NEAR	7.785 (***)	453	NEO	-0.338 (***)	13.1
QNT	0.599 (***)	12.5	SAND	1.306 (***)	30.0
SHIB	1.426 (***)	34.2	SOL	0.381 (***)	17.0
TRX	-0.188 (***)	30.2	UNI	0.107 (***)	18.0
VET	-0.316 (***)	11.2	WAVES	1.804 (***)	46.1
XLM	0.436 (***)	37.8	XMR	-0.006 (***)	16.1
XRP	1.222 (***)	57.1	XTZ	0.254 (***)	13.1
EWP	-0.853 (***)	20.5			

[a] Indicates a p-value < 0.001. [b] All kurtosis p-values were < 0.001.

tribution description of returns for all 40 cryptocurrencies, as well as the EWP. Fourteen of the cryptocurrencies, as well as the EWP, are negatively skewed. All return distributions are leptokurtic (have positive excess kurtosis). As the distributions of cryptocurrency returns are non–normal, we examine the tail behavior of the distribution (Rachev and Mittnik 2000; Shimokawa et al. 2007; Lux 2009; Rogers and Zhang 2011). Characterization of the tail statistics embodies the principle of "let the tail speak for itself" (Embrechts et al. 1997; McNeil et al. 2015). In Sections 2.2.2 and 2.2.3 we consider two methods to characterize the tails of a return distribution.

2.2.2 The Generalized Pareto Distribution

The initial approach we consider uses the generalized Pareto distribution (GPD) to model the tails of the returns. The cumulative distribution function (CDF) of the GPD is given by

$$F_{\text{GPD}}(x;\sigma,\xi) = \begin{cases} 1-(1+\xi x/\sigma)^{-1/\xi} & \text{if } \xi \neq 0\,, \\ 1-e^{-x/\sigma} & \text{if } \xi = 0\,, \end{cases} \tag{2.21}$$

where $x \in [0, \infty)$ if $\xi \geq 0$, and $x \in [0, -\sigma/\xi]$ if $\xi < 0$. Here, σ is the scale parameter and ξ is the shape parameter.[10] The sign of ξ determines the nature of the distribution tail:

- if $\xi > 0$, the distribution has a power–law decay (i.e., is heavy–tailed);
- if $\xi = 0$, the distribution decays exponentially; and
- if $\xi < 0$, the support of the distribution is bounded.

As demonstrated by Balkema and De Haan (1974) and Pickands III (1975), a broad category of tail distribution functions can be accurately approximated by the GPD.

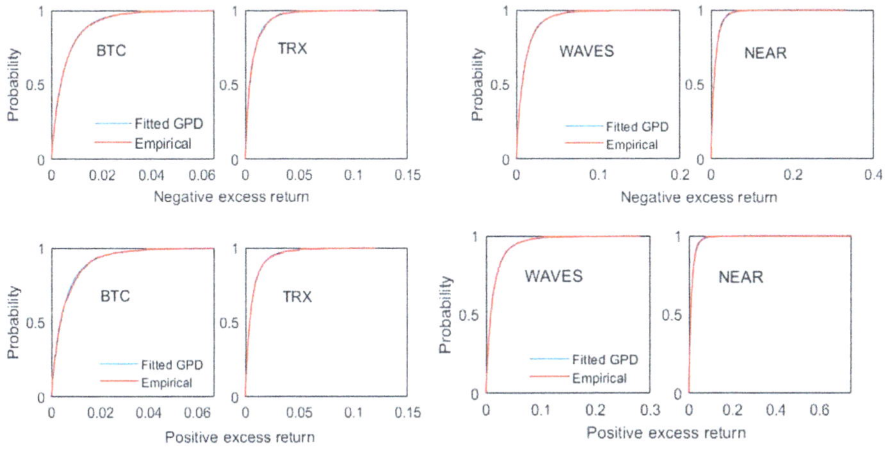

Fig. 2.9: (top) Left and (bottom) right tail returns of the empirical tail CDF and the GPD fitted CDF for the cryptocurrencies having the two lowest and two highest volatilities

We define the positive (right) tail region to consist of the largest 5% of the hourly return data for each cryptocurrency. Similarly we define the negative (left) tail region to consist of the smallest 5% of the hourly returns. As the left tail region characterizes extreme losses, a sharper focus should be placed on this tail. For completeness, we consider both tail regions. For crypto asset i, let $r_{i,5}$ and $r_{i,95}$ denote, respectively, the value of the return corresponding to the 5'th and 95'th percentiles. These correspond to the return values defining the lower and upper 5% tail regions. For values $r_i < r_{i,5}$, $|r_i - r_{i,5}|$ denotes the negative excess return; for values $r_i > r_{i,95}$, $r_i - r_{i,95}$ denotes the positive excess return. (Both excess returns are defined to be positive numbers.)

We utilized the MATLAB `gpfit()` function[11] to fit a GPD CDF to the empirical CDF of each tail region. Graphical comparison of the CDFs for the empirical data and GPD

---

10 We use the form of the GPD that assumes a location (modal) value of zero.

11 More details on the `gpfit()` function can be found at https://www.mathworks.com/help/stats/gpfit.html.

fit are shown in Fig. 2.9 for the cryptocurrencies having the two lowest and two highest volatilities. These fits are illustrative of the results for all 40 cryptocurrencies as well as for the EWP. The fits are uniformly very good, suggesting that the GPD provides a reliable model for the return tails. The function `gpfit()` returns estimates of, and 95% confidence intervals (CIs) for, the parameters σ and ξ. Graphical results for the shape parameter ξ and its 95% CI limits are presented in Fig. 2.10 for each tail for each cryptocurrency as well as for the EWP benchmark. Two results are apparent from the figure:

- the values of ξ are larger for the right tail than for the left; and
- at the 95% confidence level, the estimated values for ξ are significantly greater than zero, indicating that both tails are "heavy".

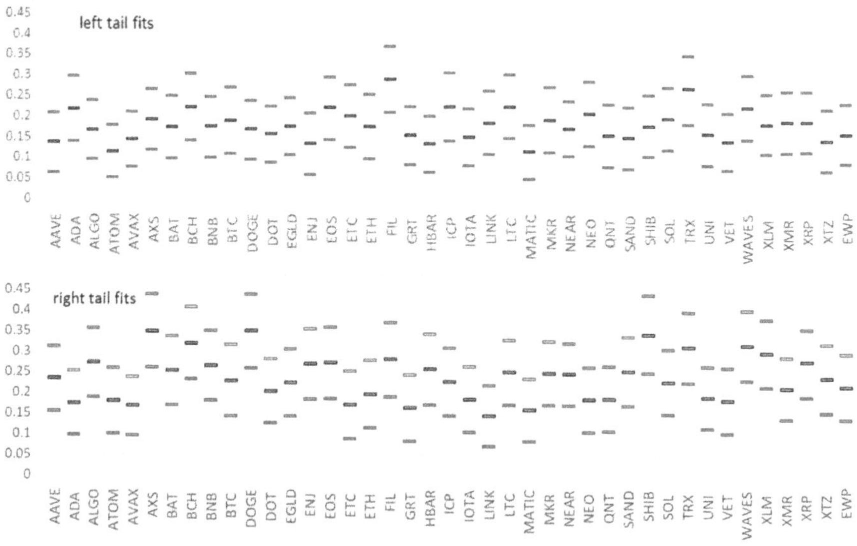

Fig. 2.10: Values for ξ (black bars) and the 95% CI limits (red bars) for the left and right tail GPD fits to the return distribution of each cryptocurrency and the EWP

2.2.3 The Hill Estimator for the Tail Index

The quantity $\alpha = 1/\xi$, where ξ is the shape parameter of the GPD, is referred to as the tail index. From (2.21) we note that the larger the value of $\alpha > 0$, the faster the power law tail decays (i.e., the faster the tail CDF rises). The Hill estimator (Hill 1975) uses sample order statistics to provide an estimate for α. Consider n independent, identically distributed (iid), positive samples, $X_1, X_2, \cdots, X_n$, from the right tail of the univariate random variable X having support on $(-\infty, \infty)$. Sort the sample values in non-increasing order, $X_{(1)} \geq X_{(2)} \geq \cdots \geq X_{(n)}$. The k'th largest value, $X_{(k)}$, is referred to as the k'th order statistic of the sample. The Hill estimator uses sample order statistics up to $k+1$ to provide the following estimate for α,

$$\widehat{\alpha}_{k,n}^{(\mathrm{H})} = \left[\frac{1}{k} \sum_{i=1}^{k} \ln X_{(i)} - \ln X_{(k+1)} \right]^{-1} . \tag{2.22}$$

The estimate has the property (De Haan and Peng 1998) that

$$\widehat{\alpha}_{k,n}^{(\mathrm{H})} \xrightarrow{\mathbb{P}} \alpha, \quad \text{as } n \to \infty,\ k \to \infty, \text{ such that } k/n \to 0 ,$$

where $\xrightarrow{\mathbb{P}}$ denotes convergence in probability. For empirical estimation, McNeil et al. (2015) recommends that the ratio k/n should be in the range $[0.01, 0.05]$. The $(1-\theta) \times$ 100% Wald CI for the estimate $\widehat{\alpha}_{k,n}^{(\mathrm{H})}$ is (Haeusler and Segers 2007)

$$\mathrm{CI}_n^{(\mathrm{Wald})}(\theta, k) = \left[\left(1 + \frac{z_{\theta/2}}{\sqrt{k}}\right)^{-1} \widehat{\alpha}_{k,n}^{(\mathrm{H})},\ \left(1 - \frac{z_{\theta/2}}{\sqrt{k}}\right)^{-1} \widehat{\alpha}_{k,n}^{(\mathrm{H})} \right],$$

where z_p denotes the critical value of the standard normal distribution corresponding to a right tail area of p.[12] We computed 95% Wald CIs ($\theta = 0.05$).

To apply the Hill estimator (2.22) to n iid, negative samples, $X_1, X_2, \cdots, X_n$, from the left tail of X, we need to consider $Y_1 = -X_1, Y_2 = -X_2, \cdots, Y_n = -X_n$, sort $Y_1, Y_2, \cdots, Y_n$ in non-increasing order, and compute (2.22) based on the Y_i values.

The GPD fits in Section 2.2.2 defined the tail region as the lowest (highest) 5% of the returns in the left (respectively, right) tail. Given a dataset of 18,495 hourly returns for each cryptocurrency, the lowest (highest) 925 returns define the left (respectively, right) tail for each asset. The recommendation of McNeil et al. (2015) would correspond to investigating the Hill estimate over order statistics $k \in [9, 46]$. However, this recommendation is very qualitative, and ultimately requires a decision made “by eye”. Consequently, we considered Hill estimates over the range $k \in [9, 90]$ ($k/n \in [0.01, 0.1]$).

Fig. 2.11 plots estimated tail indices and the 95% Wald CI as a function of k for the cryptocurrencies analyzed in Fig. 2.9. The estimated value $\widehat{\alpha}_{k,n}^{(\mathrm{H})}$ becomes “relatively” constant over the range $k \in [60, 90]$. The distance between $\widehat{\alpha}_{k,n}^{(\mathrm{H})}$ and the lower CI remains relatively constant over the entire range $k \in [9, 90]$. However the distance between the upper CI and $\widehat{\alpha}_{k,n}^{(\mathrm{H})}$ decreases rapidly over the range $k \in [9, 60]$. Based on these observations, we consider Hill estimates for $k = 65$ for the left and right tails. We further note that, over the full range $k \in [9, 90]$, the 95% CI comfortably bounds the estimated tail indices to positive values, providing further confirmation that the tails are heavy (have power law decays). These results are representative of all 40 cryptocurrencies, as well as for the EWP.

12 Equivalently, z_p is the $(1-p)$'th quantile of the standard normal distribution.

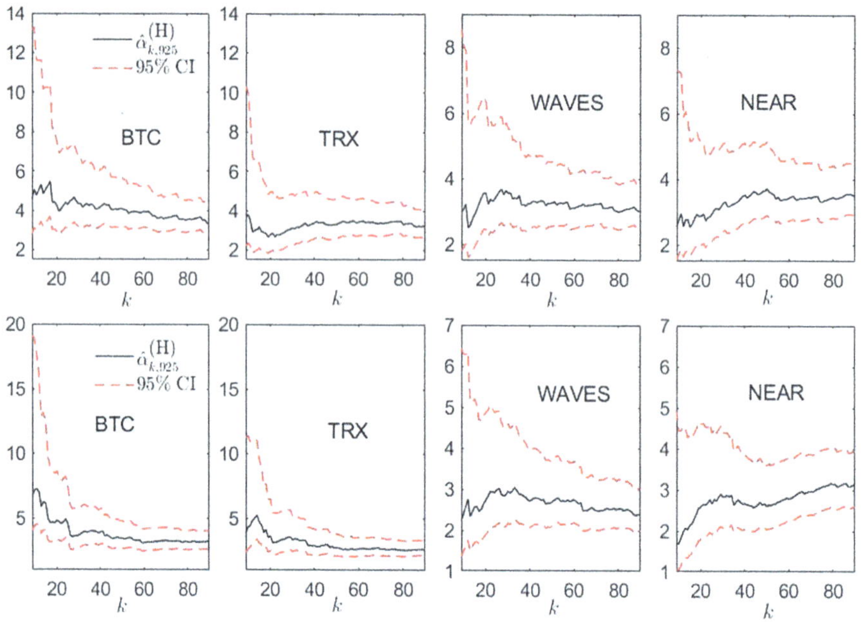

Fig. 2.11: The dependence of the estimated (top) left tail and (bottom) right tail indices $\hat{a}^{(H)}_{k,n}$, and their corresponding 95% Wald CI on the order statistic k for the cryptocurrencies having the two lowest and two highest volatilities

For each of the 40 cryptocurrencies, an estimated value $\hat{a}^{(H)}_{65,925}$ for the index a was obtained from the Hill estimator applied to their left–tail return distribution. The result is a distribution of 40 left–tail indices. Similarly, a distribution of 40 Hill estimates of right–tail indices were obtained. Analogous distributions of a were obtained from the GPD tail estimates $\overline{a} = 1/\xi$, with the ξ values given in Fig. 2.10. Fig. 2.12 presents a box–whisker summary of these distributions of a. Under both estimations, the left tail has a higher median and mean estimate than the right tail. This is accompanied by significant shifts in their inter–quartile regions. Thus the left tails of cryptocurrency (hourly) return distributions generally decay faster than their right tails. Note that the GPD estimates for the tail index produce large values than the Hill estimates.

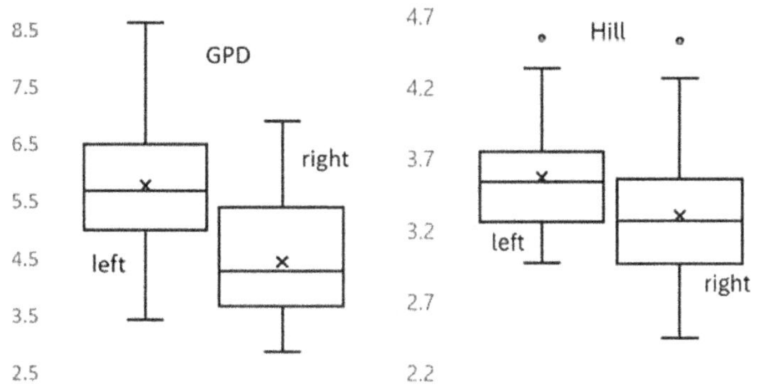

Fig. 2.12: Box–whisker summary of the left– and right–tail distributions of the cryptocurrency tail indices α obtained by the GPD estimate $\overline{\alpha} = 1/\xi$ and the Hill estimate $\widehat{\alpha}^{(H)}_{65,925}$

3 Modern Portfolio Theory

This chapter presents modern portfolio theory, focusing on two portfolio optimization techniques: the mean–variance (Markowitz 1952) and the mean–CVaR (Rockafellar and Uryasev 2000) methods. The goal of such optimizations are to minimize risk subject to attaining a desired expected return.

Section 3.1 provides the theoretical foundations of mean–variance optimization, which measures portfolio risk in terms of its standard deviation. It discusses the construction of the efficient frontier, the capital market line (CML), and the tangent portfolio, which maximizes the Sharpe ratio of a portfolio. The section demonstrates the application of mean–variance optimization to a single time–point of the empirical dataset of 40 cryptocurrencies. We end the section by presenting two alternate, equivalent formulations of mean–variance optimization. The second of these forms the basis for the robust optimization discussed in Chapter 6.

Section 3.2 introduces conditional value–at–risk (CVaR) as a coherent risk measure. As opposed to the standard deviation of mean–variance optimization, CVaR based optimization minimizes extreme tail risk, particularly important for heavy–tailed return distributions. The section outlines the theoretical foundation of CVaR–minimizing portfolios, and demonstrates the application of this optimization to a single time–point of the empirical dataset of cryptocurrencies.

3.1 Mean–Variance Portfolio Optimization

Consider a portfolio consisting of n risky assets, whose vector of returns[1] $\boldsymbol{r}(t) = [r_1(t), r_2(t), \cdots, r_n(t)]^\top$ at (historical) discrete times $t \in \{t_1, ..., t_T\}$ is considered to be a multivariate random variable having mean $\overline{\boldsymbol{r}} = (\overline{r}_1, \overline{r}_2, \cdots, \overline{r}_n)^\top = \mathbb{E}_T[\boldsymbol{r}(t)]$, where $\mathbb{E}_T[\cdot]$ stands for the expected value over the time period $[t_1, t_T]$. The returns of different assets are generally correlated, leading to an $n \times n$ covariance matrix Σ whose elements are $\Sigma_{ij} \equiv \sigma_{ij} = \mathbb{E}_T\left[(r_i(t) - \overline{r}_i)(r_j(t) - \overline{r}_j)\right]$. The diagonal elements of Σ are the variances of the individual assets, $\Sigma_{ii} = \sigma_i^2$ while $\sigma_{ij} = \sigma_{ji}$, $i \neq j$ is the covariance between assets i and j. The vector of asset standard deviations is $\boldsymbol{\sigma} = [\sigma_1, \sigma_2, \cdots, \sigma_n]^\top$.

If $\boldsymbol{w}(t_{T+1}) = [w_1(t_{T+1}), w_2(t_{T+1}), \cdots, w_n(t_{T+1})]^\top$ denotes the weights to be applied to the assets during the trading interval $(t_T, t_{T+1}]$,[2] then the expected return of the portfolio at time t_{T+1} will be given by (2.11). Because it is assumed that the portfolio is fully invested in the n risky assets, we require the condition $\sum_{i=1}^{n} w_i(t_{T+1}) = 1$.[3] As the as-

1 As noted previously, we consider arithmetic returns.

2 Recall, the weights are determined at t_T and are assumed not to change during the trading interval $(t_T, t_{T+1}]$.

3 In vector notation, $\boldsymbol{e}_n^\top \boldsymbol{w}(t_{T+1}) = 1$, where $\boldsymbol{e}_n = [1, 1, \cdots, 1]^\top \in \mathbb{R}^n$.

https://doi.org/10.1515/9781501517136-003

set returns are random variables, the portfolio expected return $r_p(t_{T+1})$ is also a random variable having mean value $\bar{r}_p(t_{T+1}) = \boldsymbol{w}^\top(t_{T+1})\bar{\boldsymbol{r}} = \bar{\boldsymbol{r}}^\top \boldsymbol{w}(t_{T+1})$ and variance $\sigma_p^2(t_{T+1}) = \boldsymbol{w}^\top(t_{T+1})\Sigma\boldsymbol{w}(t_{T+1})$.[4] In the following presentation, we consider optimization at an arbitrary time value t_T, based on information available over the previous time period $t \in \{t_1, ..., t_T\}$. For notational simplicity, we therefore drop the time notation.

The objective of portfolio optimization using the method of Markowitz (1952) is to determine the set of asset weights for the next time period that minimize the portfolio's risk while meeting a desired expected (i.e., mean) return $\bar{r}_p$. The choice of $\bar{r}_p$ reflects the investor's level of risk aversion – a higher expected return reflects a greater willingness to take on risk. The Markowitz optimization uses the portfolio standard deviation σ_p as the measure of portfolio risk. Specifically mean (return)–variance optimization is the minimization of the variance

$$\min_{\boldsymbol{w}} \boldsymbol{w}^\top\Sigma\boldsymbol{w} \quad \text{subject to} \begin{cases} \bar{\boldsymbol{r}}^\top \boldsymbol{w} = \bar{r}_p \,, \\ \boldsymbol{e}_n^\top \boldsymbol{w} = 1 \,. \end{cases} \tag{3.1}$$

As the desired return $\bar{r}_p$ is varied, the optimum solution $(\sigma_p, \bar{r}_p)$ traces out a curve known as the *portfolio frontier*.

The optimization problem (3.1) can be solved via the standard method of Lagrange multipliers. Solving (3.1) is equivalent to solving[5]

$$\min_{\boldsymbol{w},q,\theta_0} L(\boldsymbol{w}, q, \theta_0) = \min_{\boldsymbol{w},q,\theta_0} \left[\frac{\boldsymbol{w}^\top\Sigma\boldsymbol{w}}{2} + q(\bar{r}_p - \bar{\boldsymbol{r}}^\top\boldsymbol{w}) + \theta_0(1 - \boldsymbol{e}_n^\top\boldsymbol{w})\right]. \tag{3.2}$$

The variables q and θ_0 are Lagrange multipliers used to enforce the respective constraints on $\bar{\boldsymbol{r}}^\top\boldsymbol{w}$ and $\boldsymbol{e}_n^\top\boldsymbol{w}$ in (3.1). Applying the first–order optimizing conditions[6]

$$\begin{aligned} \partial L(\boldsymbol{w}, q, \theta_0)/\partial w_i &= 0\,, \quad i = 1, 2, \cdots, n\,, \\ \partial L(\boldsymbol{w}, q, \theta_0)/\partial q &= 0\,, \\ \partial L(\boldsymbol{w}, q, \theta_0)/\partial \theta_0 &= 0\,, \end{aligned}$$

(3.2) yields the following system of linear equations,

$$\Sigma\boldsymbol{w} - q\bar{\boldsymbol{r}} - \theta_0\boldsymbol{e}_n = 0\,, \quad \bar{\boldsymbol{r}}^\top\boldsymbol{w} = \bar{r}_p\,, \quad \boldsymbol{e}_n^\top\boldsymbol{w} = 1\,. \tag{3.3}$$

4 Given the trivial relationship $\sigma_p(t_{T+1}) = \sqrt{\sigma_p^2(t_{T+1})}$ between the standard deviation and the variance of a portfolio, minimizing the variance is equivalent to minimizing the standard deviation. As $\bar{r}_p(t_{T+1})$ and $\sigma_p(t_{T+1})$ have the same units, in mean–variance optimization it is natural to express the risk in terms of the portfolio standard deviation. However, being quadratic in form, the variance is more natural to use in the optimization formulation (3.1).

5 As $\boldsymbol{w}^\top\Sigma\boldsymbol{w}$ is quadratic, the minimizing values of $\boldsymbol{w}^\top\Sigma\boldsymbol{w}$ and of $\boldsymbol{w}^\top\Sigma\boldsymbol{w}/2$ are identical. The choice $\boldsymbol{w}^\top\Sigma\boldsymbol{w}/2$ is used to simplify the result of the partial derivatives $\partial L/\partial w_i$, $i = 1, ..., n$.

6 As $L(\boldsymbol{w}, q, \theta_0)$ is quadratic in $\boldsymbol{w}$ with a positive coefficient multiplying the quadratic term, the first–order optimizing conditions determine a global minimum.

The analytic solution to the system (3.3) is

$$w^* = \bar{r}_p w_1 + w_2 , \tag{3.4}$$

where

$$\begin{aligned} w_1 &= \frac{1}{\Delta}\left(B\Sigma^{-1}\bar{r} - C\Sigma^{-1}e_n\right), \quad w_2 = \frac{1}{\Delta}\left(A\Sigma^{-1}e_n - C\Sigma^{-1}\bar{r}\right), \\ \Delta &= AB - C^2, \quad A = \bar{r}^\top\Sigma^{-1}\bar{r}, \quad B = e_n^\top\Sigma^{-1}e_n, \quad C = \bar{r}^\top\Sigma^{-1}e_n . \end{aligned} \tag{3.5}$$

Solving for σ_p^* gives

$$\sigma_p^* = \sqrt{w^{*\top}\Sigma w^*} = \sqrt{\frac{B\bar{r}_p^2 - 2C\bar{r}_p + A}{\Delta}} . \tag{3.6}$$

From (3.6), we see that the portfolio frontier points $(\sigma_p^*, \bar{r}_p)$ trace out a hyperbola in (standard deviation, return) space. This hyperbola has a minimum at the point (i.e. the portfolio) $(\sigma_{p,\min}, \bar{r}_{p,\min})$ corresponding to the value $w^*_{\min} = \Sigma^{-1}e_n/B$. This specific portfolio, having expected return $\bar{r}_{p,\min} = C/B$ and standard deviation $\sigma_{p,\min} = 1/\sqrt{B}$, is known as the *minimum–variance* portfolio. Portfolios on the frontier with expected mean return $\geq \bar{r}_{p,\min}$ are said to lie on the *efficient frontier*.[7]

In addition to the minimum–variance portfolio, which determines the “low–risk” end point of the efficient frontier, there is a second portfolio that determines the “high–risk” end point. The existence of this end point is based on the fact that if $\bar{r}_p$ is chosen too large, then no solution to the optimization problem (3.1) is possible. A non–rigorous, intuitive argument for the existence of the high–risk limit goes as follows.[8] Given the two constraints in (3.1), then $\bar{r}_p \geq \bar{r}_k = \max_{i=1,\dots,n}(\bar{r}_i)$. Achieving an expected return larger than $\bar{r}_k$ requires $\sum_{i=1}^n w_i\bar{r}_i > \bar{r}_k$, for some set of weights $\{w_i\}$. For a given set of values $\{\bar{r}_i\}$ and the restriction $\sum_{i=1}^n w_i = 1$, this is may not be possible. Consequently, the highest risk portfolio on the efficient frontier has $\bar{r}_p = \bar{r}_k$ and $\sigma_p = \sigma_k$, with weights given by $w_k = 1$, $w_i = 0$, $i = 1, \dots, n$, $i \neq k$. This intuitive argument does not take into account more complex constraints. MATLAB provides the function `estimateFrontierLimits()` to determine the efficient frontier endpoints based upon an optimization problem with a given set of constraints.

Additional constraints can be added to the minimization problem. For example, the minimization

$$\begin{aligned} \min_{w,q,\theta_0,\boldsymbol{\theta}_1,\boldsymbol{\theta}_2} L(w, q, \theta_0, \boldsymbol{\theta}_1, \boldsymbol{\theta}_2) = \min_{w,q,\theta_0,\boldsymbol{\theta}_1,\boldsymbol{\theta}_2} \Bigg[&\frac{w^\top\Sigma w}{2} + q\left(\bar{r}_p - \bar{r}^\top w\right) \\ &+ \theta_0\left(1 - e_n^\top w + \boldsymbol{\theta}_1^\top (w_{\text{lb}} - w) + \boldsymbol{\theta}_2^\top (w - w_{\text{ub}})\right)\Bigg] , \end{aligned} \tag{3.7}$$

7 If a portfolio lies on the mean–variance efficient frontier, no other portfolio (fully invested in the same n risky assets) can exist that has the same standard deviation but a higher expected return. Equivalently, no other portfolio can exist that has the same expected return but a lower standard deviation.

8 In practice, the argument holds surprisingly often. See e.g., Figs. 3.1 and 3.3.

imposes two additional constraints, the weights must satisfy $w_{\mathrm{lb},i} \le w_i \le w_{\mathrm{ub},i}$, for $i = 1, 2, \cdots, n$. The elements of the vectors $\boldsymbol{\theta}_1$ and $\boldsymbol{\theta}_2$ are the Lagrange multipliers[9] enforcing these bound constraints. The first–order optimizing conditions,

$$\begin{aligned}
\partial L(\mathbf{w}, q, \theta_0, \boldsymbol{\theta}_1, \boldsymbol{\theta}_2)/\partial w_i &= 0\,, \quad i = 1, 2, \cdots, n\,,\\
\partial L(\mathbf{w}, q, \theta_0, \boldsymbol{\theta}_1, \boldsymbol{\theta}_2)/\partial q &= 0\,,\\
\partial L(\mathbf{w}, q, \theta_0, \boldsymbol{\theta}_1, \boldsymbol{\theta}_2)/\partial \theta_0 &= 0\,,\\
\partial L(\mathbf{w}, q, \theta_0, \boldsymbol{\theta}_1, \boldsymbol{\theta}_2)/\partial \theta_{1,i} &= 0\,, \quad i = 1, 2, \cdots, n\,,\\
\partial L(\mathbf{w}, q, \theta_0, \boldsymbol{\theta}_1, \boldsymbol{\theta}_2)/\partial \theta_{2,i} &= 0\,, \quad i = 1, 2, \cdots, n\,,
\end{aligned}$$

provide sufficient conditions for a numerical solution of the values $\mathbf{w}, \theta_0, \boldsymbol{\theta}_1$ and $\boldsymbol{\theta}_2$. Equation (3.7) defines a, so–called, long–only strategy by setting $w_{\mathrm{lb}} = 0$ and $\boldsymbol{\theta}_2 = 0$[10] or a long–short strategy by assigning appropriate negative values to some entries of w_{lb} while keeping $\boldsymbol{\theta}_2 = 0$.

From these two optimization examples, it is clear the precise shape of the efficient frontier is influenced not only by the risky assets (their number, mean return $\overline{\mathbf{r}}$, and covariance structure Σ) but also by the constraints imposed. For example, mean–variance efficient frontiers computed numerically under constraints on the percentage holding of any single asset can be found in Bloch et al. (1993). When performing practical portfolio optimization, constraints must control the cost of maintaining the portfolio (buy and sell transaction costs, margin costs for long–short strategies), consideration of asset availability through brokers, and many other factors. Key constraints may include: available budget; holding period (length of time $\mathbf{w}$ is left unchanged); control of a variety of risk factors; transaction size (allowed amounts of asset purchase or sale); cardinality (number of assets); volatility of the portfolio; turnover (total allowed changes in $\mathbf{w}$); and tracking error (how much the portfolio return deviates from that of a benchmark, such as an index, it is intended to track). In addition to the expected return and weight constraints illustrated in (3.7), in Chapter 4 we consider holding period and turnover constraints. Discussion of other constraints is beyond the scope of this book.

In addition to affecting the shape of the efficient frontier, the set of constraints can be so restrictive as to, at certain times, result in no possible solution to the optimization problem (i.e., an empty feasible set). A common procedure to guard against such occurrences is to incorporate the constraint into the objective function of the optimization problem using a penalty term. Such a procedure allows the constraint to be violated, with the penalty function controlling the size of the violation. Such so–called "soft–constraint" methods are also beyond the scope of this book. In the constrained optimization problems considered here, if the feasible set is empty at a time t, the weights

9 Specifically, they are $\boldsymbol{\theta}_1^\top = (\theta_{1,1}, \theta_{1,2}, \cdots, \theta_{1,n})$ and $\boldsymbol{\theta}_2^\top = (\theta_{2,1}, \theta_{2,2}, \cdots, \theta_{2,n})$.

10 A long–only strategy is characterized by bounds $0 \le w_i \le 1$, $i = 1, \ldots, n$. However, it unnecessary to specify $w_{lb} = 1$, as the requirements $w_i \ge 0$, $i = 1, 2, \cdots, n$ and $\mathbf{e}_n^\top \mathbf{w} = 1$ are sufficient.

used for the previous time interval $(t-\Delta t, t]$ are also used for the time interval $(t, t+\Delta t]$. (However, see Chapter 5 where a sequence of relaxed turnover constraint values is used in the case of an empty feasible set.)

3.1.1 Capital Market Line and the Mean–Variance Tangent Portfolio

We consider the effect of adding a single riskless asset having the guaranteed return r_f to a portfolio of n risky assets. Denoting the weight of the riskless asset in the portfolio by w_f, we have the fully–invested condition $\boldsymbol{e}_n^\top \boldsymbol{w} + w_f = 1$. The mean–variance optimization problem is now

$$\min_{\boldsymbol{w}} \frac{\boldsymbol{w}^\top \Sigma \boldsymbol{w}}{2} \quad \text{subject to} \quad \begin{cases} \boldsymbol{e}_n^\top \boldsymbol{w} + w_f = 1 \,, \\ w_f r_f + \overline{\boldsymbol{r}}^\top \boldsymbol{w} = \overline{r}_p \,. \end{cases} \tag{3.8}$$

By eliminating w_f, the two constraints in (3.8) can be combined into a single constraint and (3.8) rewritten as

$$\min_{\boldsymbol{w}} \frac{\boldsymbol{w}^\top \Sigma \boldsymbol{w}}{2} \quad \text{subject to} \ (1 - \boldsymbol{e}_n^\top \boldsymbol{w}) r_f + \overline{\boldsymbol{r}}^\top \boldsymbol{w} = \overline{r}_p \,. \tag{3.9}$$

The equivalent Lagrange multiplier optimization problem is

$$\min_{\boldsymbol{w},q} L(\boldsymbol{w}, q) = \min_{\boldsymbol{w},q} \left\{ \frac{\boldsymbol{w}^\top \Sigma \boldsymbol{w}}{2} + q \left[\overline{r}_p - (1 - \boldsymbol{e}_n^\top \boldsymbol{w})\, r_f - \overline{\boldsymbol{r}}^\top \boldsymbol{w} \right] \right\} . \tag{3.10}$$

Again, the desired expected return $\overline{r}_p$ is an investor–specified parameter. The first–order optimizing conditions are

$$\begin{aligned} \partial L(\boldsymbol{w}, q)/\partial w_i &= 0 \,, \quad i = 1, 2, \cdots, n \,, \\ \partial L(\boldsymbol{w}, q)/\partial q &= 0 \,, \end{aligned}$$

having solution

$$\boldsymbol{w} = q(\overline{r}_p) \Sigma^{-1} (\overline{\boldsymbol{r}} - r_f \boldsymbol{e}_n) \,, \qquad q(\overline{r}_p) = \frac{\overline{r}_p - r_f}{B r_f^2 - 2 C r_f + A} \,, \tag{3.11}$$

where A, B, and C are given in (3.5). The portfolio standard deviation σ_p is

$$\sigma_p = \frac{1}{\sqrt{B r_f^2 - 2 C r_f + A}} \overline{r}_p - \frac{r_f}{\sqrt{B r_f^2 - 2 C r_f + A}} \,, \tag{3.12}$$

where (3.12) is written to emphasize that σ_p varies linearly with $\overline{r}_p$. The points $(\sigma_p, \overline{r}_p)$ of the portfolio frontier now trace out a straight line, the *capital market line* (CML).

When $w_f = 1$ (i.e., $\bar{r}_p = r_f$), the CML intersects the expected return axis at $(0, r_f)$; when $w_f = 0$ (i.e., $\boldsymbol{e}_n^\top \boldsymbol{w} = 1$), the CML meets the efficient frontier[11] tangentially[12] at the *market portfolio* value $(\sigma_m, \bar{r}_m)$. The weights $\boldsymbol{w}_m$ for the market portfolio are determined from (3.11),

$$\boldsymbol{w}_m = \frac{\Sigma^{-1}(\bar{\boldsymbol{r}} - r_f \boldsymbol{e}_n)}{C - r_f B} . \tag{3.13}$$

The mean return of the market portfolio is $\bar{r}_m = \bar{\boldsymbol{r}}^\top \boldsymbol{w}_m$. Its standard deviation, σ_m, is given by (3.12) when $\bar{r}_p = \bar{r}_m$. Evaluating these quantities yields

$$\sigma_m = \frac{\sqrt{Br_f^2 - 2Cr_f + A}}{C - r_f B} , \qquad \bar{r}_m = r_f + \frac{Br_f^2 - 2Cr_f + A}{C - r_f B} . \tag{3.14}$$

From (3.14), we have the identity $(\bar{r}_m - r_f)/\sigma_m = \sqrt{Br_f^2 - 2Cr_f + A}$. This identity can be used to rewrite (3.12) for the CML as

$$\bar{r}_p = r_f + \frac{\bar{r}_m - r_f}{\sigma_m} \sigma_p . \tag{3.15}$$

From (3.15), we note

$$\frac{\bar{r}_p - r_f}{\sigma_p} = \frac{\bar{r}_m - r_f}{\sigma_m} , \tag{3.16}$$

verifying that all the portfolios $(\sigma_p, \bar{r}_p)$ on the CML have the same Sharpe ratio (Sharpe 1966).[13] Equation (3.15) can be written in the form of the capital asset pricing model (CAPM)

$$\bar{r}_p - r_f = \beta_p(\bar{r}_m - r_f),$$

where $\beta_p = \sigma_p/\sigma_m$ denotes the CML sensitivity[14] of the expected excess return of portfolio p relative to the expected excess return of the market portfolio m.

The market portfolio $(\sigma_m, \bar{r}_m)$ is also known as the *tangent* mean–variance portfolio. Of all possible portfolios on the efficient frontier of portfolios composed solely of the n risky assets, the tangent portfolio has the largest Sharpe ratio.

We note that the CML can be extended beyond the market portfolio (i.e., setting $\boldsymbol{e}_n^\top \boldsymbol{w} > 1$) to higher expected return values carrying higher risk. Doing so requires $w_f < 0$, i.e., leveraging a position by borrowing at the risk–free rate to invest additional capital into the portfolio. Successful leveraging anticipates that excess returns above $\bar{r}_m$ will cover the interest required to carry the loan. Consideration of such a strategy must weigh the potential for higher return against the increased exposure to market uncertainty.

11 Here we mean the efficient frontier of optimized portfolios containing only the n risky assets.

12 It takes some calculation to show that the portfolio point (3.14) lies on the efficient frontier hyperbola (3.6). Similarly, the slope of the efficient frontier (3.6) at the point $(\sigma_m, \bar{r}_m)$ can be shown to match the slope of the CML.

13 See (4.10) for the definition of the Sharpe ratio.

14 Commonly referred to as "beta" in Wall Street terminology.

3.1.2 Empirical Application of Mean–Variance Optimization using MATLAB

We demonstrate the application of mean–variance optimization to a portfolio composed of the cryptocurrency dataset introduced in Chapter 1. To aid the understanding of this implementation, we demonstrate using MATLAB code.[15] The objective of this code is the following. Given historical data available at 05:00:00 on September 8, 2023, compute optimized portfolio weights to be used during the next time period 05:00:00 through 06:00:00. The first code block loads the price data and computes arithmetic returns. Extensive comments in the code listing describe the action of each function call.

Listing 3.1: Compute arithmetic returns

```
% Load the price dataset from a text file containing n+1 columns with
% timestamps in column 1 and cryptocurrency prices in columns 2 through
% n+1. The header row of the text file consists of column names.
pr_data = readtable('Price_1h.txt');

% Store cryptocurrency names in a row vector.
cr_nms = pr_data.Properties.VariableNames(2:end);

% Load prices into a matrix, required as input for price2ret()
prices = pr_data(:,2:end);
prices = table2array(prices);

% Compute returns for every cryptocurrency. The Method character string
% 'periodic' denotes computation of arithmetic prices; 'continuous'
% denotes log-returns.
ret = price2ret(prices, Method='periodic');

% Practice good memory maintenance. There is no need to maintain the
% price data once the return data are computed.
clear prices;
```

Mean–variance optimization requires the mean returns for each asset as well as the covariance matrix of the asset returns. The next code segment performs this task.

Listing 3.2: Compute mean and covariance

```
% Compute the vector of asset mean returns and the covariance matrix of
% the asset returns.
mu = mean(ret);  % mu is a column n-vector
```

15 Python, R and C++ are also commonly used in finance computations. Hilpisch (2018) provides a readable introduction to the use of Python.

```
mu = mu';     % mu is converted to a row vector
Sigma = cov(ret);  % Sigma is an n x n matrix
```

As the optimization provides weights to be applied for the next trading hour, computation of the tangent portfolio requires a current hourly value of the risk-free rate. We utilized daily yield rates for US three-month treasury bills, which are provided as equivalent annual percentages. The rates are released at 15:30 Eastern time each business day of the federal government. For example, the three month treasury rate released at 15:30 on 04/26/2022 was $r_{3\text{Mo}} = 0.83\%$, annualized. This is equivalent to a constant, arithmetic, hourly rate of

$$r_{\text{hr}} = (1 + 0.0083)^{1/(365\times24)} - 1 \,.$$

This hourly rate was then used for the following 24 hourly optimizations (until the next three month rate was released). However, the timestamps associated with the cryptocurrency data is UTC+0 whereas U.S. Eastern standard time is UTC-05:00 and U.S. Eastern daylight savings is UTC-04:00.[16] Thus, 15:30 EST corresponds to 20:30 UTC+0; while 15:30 EDT corresponds to 19:30 UTC+0. To continue our example, the treasury rate released on 04/26/2022 (when the U.S. was on daylight savings) would be applied to optimizations having UTC+0 timestamps from from 20:00 on 04/26/2022 to 19:00 on 04/27/2022. As cryptocurrency trading is 24/7 while treasury rates are released on federal business days, obvious modifications are required for weekends and federal holidays. (For example, the three–month rate released on 04/29/2022 would be used for optimizations covering the hourly periods from 20:00 04/29/2022 through 19:00 05/02/2022.)

The next code block finds the risk–free rate appropriate for the optimization to be performed at 05:00:00 on September 8, 2023.

Listing 3.3: Set risk–free rate

```
% Import daily treasury yields. The input csv file is column-oriented
% having the column entries: Date, 1 Mo, 2 Mo, 3 Mo, ..., 30 yr.
% MATLAB's readtable() converts these to: Date, x1Mo, x2Mo, ..., x30Yr.

% Import rf data table. Specification of the date format is done below.
rf_table = readtable('daily-tr-2021-23.csv');

% Store 3 Mo yields in a column vector and convert to hourly arithmetic
% rates.
rf_3mo = rf_table.x3Mo;
rf_3mo = rf_3mo ./ 100; % Convert to annual fraction.
nhr = 1 / (365 * 24); % There are no leap years in our data set.
```

16 The MATLAB `datetime` functionality is capable of conversion between time zones and takes into account the standard time to daylight savings time changeovers in every time zone country.

```
rf_3mo = (1 + rf_3mo) .^ nhr - 1; % Convert to hourly.

% Store rf dates in a DateTime vector.
% The daily dates contain no hr, min, or sec data.
rf_times = rf_table.Date;

% Add hr, min, sec and time zone format to rf dates.
% 'HH', specifies a 24 hour clock.
rft_NY = datetime(rf_times,'InputFormat','MM/dd/yyyy', ...
                 'Format','MM/dd/yyyy HH:mm:ss Z', ...
                 'TimeZone','America/New_York');

% Set hr, min, sec to 15:30:00 NY time.
t1 = duration([15 30 00]);
rft_NY = rft_NY + t1;

% Convert NY time to Greenwich Mean Time (UTC+0).
rft_UTC = rft_NY;
rft_UTC.TimeZone = 'Etc/GMT';

% Clear unneeded memory.
clear rf_table rft_times rft_NY t1;

% Store crypto return times in a DateTime column vector.
% No time zone has yet been specified for the crypto times.
rtn_times = pr_data.time(2:end,1);

% Reset datetime format and specify the time zone as GMT (UTC+0).
rtn_UTC = datetime(table2array(rtn_times), ...
                      'InputFormat','dd-MM-yyyy HH:mm:ss', ...
                      'Format','MM/dd/yyyy HH:mm:ss Z',...
                      'TimeZone','Etc/GMT');

clear rtn_times;

% Find the risk-free rate appropriate for the optimization to be
% performed at 05:00:00 on 09/08/2023.
Nrtn = length(rtn_UTC);
rtn_t = rtn_UTC(Nrtn,1); % corresponds to 05:00:00 on 09/08/2023

% Loop (in reverse order) over the risk-free datetimes to find the
% appropriate time and its corresponding rate.
```

```
Nrf = length(rft_UTC);
jt = Nrf;
rf_t = rft_UTC(jt,1);
while rf_t > rtn_t
        jt = jt-1;
        rf_t = rft_UTC(jt,1);
end
rf_rate = rf_3mo(jt,1);
```

In MATLAB, a mean–variance portfolio optimization problem is specified via a structure (which MATLAB refers to as an "object"), which has entries[17] that specify the number and names of the assets in the portfolio, their mean returns and covariance matrix, the appropriate risk–free rate (whose value must be updated appropriately each optimization time) as well as a set of possible constraints to be applied. The MATLAB function `Portfolio()` initializes a structure of default values, which can be reset to user specification. There are three ways to specify each structure value;[18] the code block below illustrates each of these methods.

The next code block establishes a mean–variance portfolio structure for the optimization (3.7) subject to a long–only strategy.

Listing 3.4: Establish the mean–variance portfolio object

```
% Create a mean-variance portfolio object; set the mean and
% covariance entries and the asset names.
p = Portfolio('AssetMean', mu, 'AssetCovar', Sigma); % method one
p = p.setAssetList(cr_nms);  % method two

% Set the appropriate risk free rate for 05:00:00 on September 8, 2023.
p.RiskFreeRate = rf_rate; % method three

% Set default constraints. These constrain the weights to be
% non-negative and to sum to unity,
% i.e., p.LowerBound = 0, p.LowerBudget = p.UpperBudget = 1.
p = setDefaultConstraints(p);

% As the optimization is a quadratic programming problem, specify the
% MATLAB quadprog() solver.
p = setSolver(p,'quadprog');
```

17 See https://www.mathworks.com/help/finance/portfolio.html for details.

18 Through the `'Name',Value` convention in the argument list of the `Portfolio` function, `p = Portfolio(..., 'Name',Value, ...)`; via an explicit function call of the generic form `p = p.setValue(val)`; or directly using the function structure name, having the generic form `p.Value = val`.

The next code block computes the efficient frontier at 05:00:00 on September 8, 2023, as well as the CML, the minimum risk portfolio and the tangent portfolio. Any choice of w^* on this efficient frontier (or on the CML if a riskless asset is included) determines the weights applied to the portfolio for the following hour, from 05:00:00 to 06:00:00 on September 8, 2023.

Listing 3.5: Compute mean–variance efficient frontier

```
% Estimate portfolio weights on the efficient frontier using 50 points.
% 50 points are chosen as the efficient frontier is to be plotted below.
% If no plots are required, the global mimimum portfolio can be found by
% the command
% EF_Wgts = estimateFrontier(p, 1);
% at a considerable savings in execution time.
EF_Wgts = estimateFrontier(p, 50);

% Compute [sigma_p, r_p] for all points. Note that
% [EF_Risks(1,1), EF_Rtns(1,1)] is the global minimum portfolio.
[EF_Risks, EF_Rtns] = estimatePortMoments(p, EF_Wgts);

% Estimate the tangent portfolio using the appropriate MATLAB
% optimization routine.
[Tan_Wgts,~,~] = estimateMaxSharpeRatio(p,'Method','direct');
[Tan_Risk, Tan_Rtn] = estimatePortMoments(p, Tan_Wgts);
```

The final code block plots the efficient frontier.[19] It plots the CML, identifies the tangent portfolio, and plots the points $(\sigma_i, \bar{r}_i)$, $i = 1, \ldots, n$, for each individual asset.

Listing 3.6: Plot efficient frontier

```
pfig = figure();

% Plot the efficient frontier as a black solid line.
plot(EF_Risks, EF_Rtns, 'k-', 'LineWidth', 2);
hold on;

% Plot the capital market line as a red dashed line.
x = [0 Tan_Risk];
y = [rf_rate Tan_Rtn];
plot(x, y, 'r--', 'LineWidth', 2);
hold on;
```

19 MATLAB provides a separate function, `plotFrontier()`, which only plots the efficient frontier without the additional features provided by Listing 3.6.

```
% Identify the tangent portfolio.
plot(Tan_Risk,Tan_Rtn,'ro','MarkerSize',8);
hold on;

% Plot individual assets.
mu = mu'; % convert back to column vector
sigma = std(ret); % standard deviations of asset returns.
scatter(sigma, mu, 'filled');

% Specify plot screen position and length/width (in points).
set(pfig,"Position",[50,250,750,500]);

% Specify default fontsize, axes labels and legend
fontsize(pfig,14,'points');
xlabel('Standard Deviation,'Fontsize',24);
ylabel('Expected Return', 'Fontsize',24);
legend({'Eff. Front.','CML','Tangent'},'Location','northwest');

% add grid lines.
grid on;
```

The resultant plot is displayed in Fig. 3.1. Efficient frontier portfolios vary from the minimum risk portfolio $(0.006, 1.03 \cdot 10^{-5})$ to the maximum endpoint $(0.013, 9.7 \cdot 10^{-5})$. The tangent portfolio $(0.0093, 7.6 \cdot 10^{-5})$ has an hourly Sharpe ratio of $7.5 \cdot 10^{-3}$. The individual cryptocurrencies are represented in the plot by the points $(\sigma_i, \overline{r}_i)$. An arrow points to the asset (SHIB) that, in this case, does determine the upper limit of the efficient frontier.

With this first example, we note the following practical considerations. This optimization computation must be performed "blindingly" fast, as buy and sell orders must then be executed in order to rebalance the portfolio to the new set of optimized weights. The optimization and rebalancing must be totally automated in a manner to be completed in fractions of a minute, in order to have the portfolio invested over most of the next full hour.

3.1.3 Alternate Equivalent Formulations of Mean–Variance Optimization

Fig. 3.2 illustrates that a portfolio on the efficient frontier satisfies three equivalent optimization problems. The first, denoted min $\sigma_p^2|_{r_p}$, is the optimization problem (3.1). The second, denoted max $r_p|_{\sigma_p^2}$, is the optimization

$$\max_{\boldsymbol{w}} \overline{\boldsymbol{r}}^{\top} \boldsymbol{w}, \quad \text{subject to} \begin{cases} \boldsymbol{w}^{\top} \Sigma \boldsymbol{w} = \sigma_p^2 \,, \\ \boldsymbol{e}_n^{\top} \boldsymbol{w} = 1 \,. \end{cases} \tag{3.17}$$

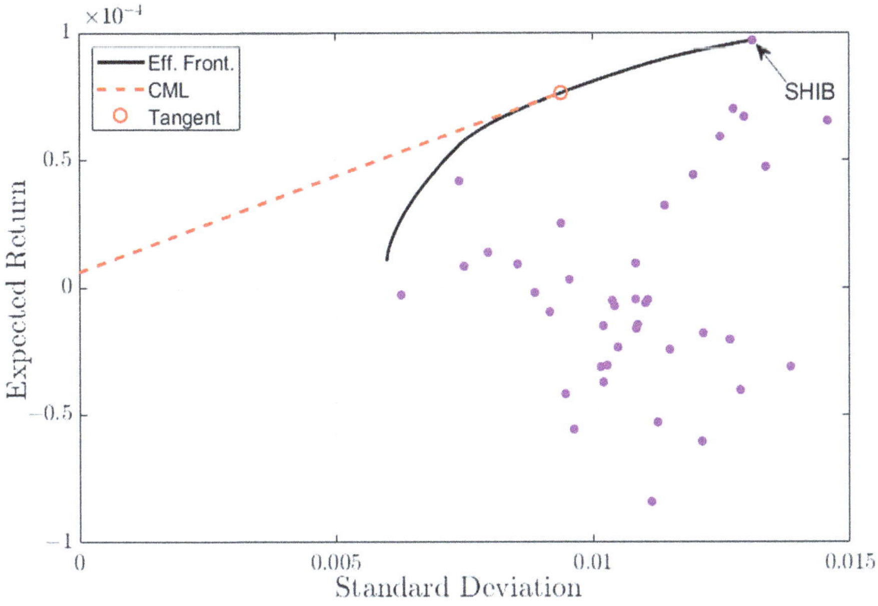

Fig. 3.1: Cryptocurrency mean–variance portfolio efficient frontier on 05:00:00, September 8, 2023 (Eff. Front.: efficient frontier; CML: capital market line; Tangent: tangent portfolio. Filled circles correspond to historical average return and standard deviation of individual crypto assets)

In the third optimization, denoted $\max\left(\beta r_p - (1-\beta)\hat{\sigma}_p^2\right)|_\beta$, the term $\hat{\sigma}_p^2 = \sigma_p^2 - \sigma_p^{2,\min}$, where $\sigma_p^{2,\min}$ is the global minimum–variance of the efficient frontier. This can be written as the optimization problem,

$$\max_{\boldsymbol{w}}\left(\beta\bar{\boldsymbol{r}}^\top\boldsymbol{w} - (1-\beta)\boldsymbol{w}^\top\Sigma\boldsymbol{w} + (1-\beta)\sigma_p^{2,\min}\right), \quad \text{subject to } \boldsymbol{e}_n^\top\boldsymbol{w} = 1\,, \tag{3.18}$$

for a fixed value of $\beta \in [0,1]$, which represents risk aversity of the portfolio manager. (If $\beta = 1$, the manager desires to maximize return regardless of the risk. As β decreases, the manager is willing to forgo return in order to lessen risk.) As $\sigma_p^{2,\min}$ is constant under this optimization, it can be written

$$\max_{\boldsymbol{w}}\left(\beta\bar{\boldsymbol{r}}^\top\boldsymbol{w} - (1-\beta)\boldsymbol{w}^\top\Sigma\boldsymbol{w}\right), \quad \text{subject to } \boldsymbol{e}_n^\top\boldsymbol{w} = 1\,. \tag{3.19}$$

This can be expressed instead as the minimization

$$\min_{\boldsymbol{w}}\left(-\beta\bar{\boldsymbol{r}}^\top\boldsymbol{w} + (1-\beta)\boldsymbol{w}^\top\Sigma\boldsymbol{w}\right), \quad \text{subject to } \boldsymbol{e}_n^\top\boldsymbol{w} = 1\,, \tag{3.20}$$

where $-\bar{\boldsymbol{r}}^\top\boldsymbol{w}$ is the loss function for the portfolio. The optimizations (3.17) and (3.20) can be solved using the same Lagrangian technique as (3.2)–(3.4). Writing the Lagrangian for (3.17) as

$$L(\boldsymbol{w}, q_1, \theta_1) = \bar{\boldsymbol{r}}^\top\boldsymbol{w} + q_1\left(\frac{\sigma_p^2}{2} - \frac{\boldsymbol{w}^\top\Sigma\boldsymbol{w}}{2}\right) + \theta_1(1 - \boldsymbol{e}_n^\top\boldsymbol{w}),$$

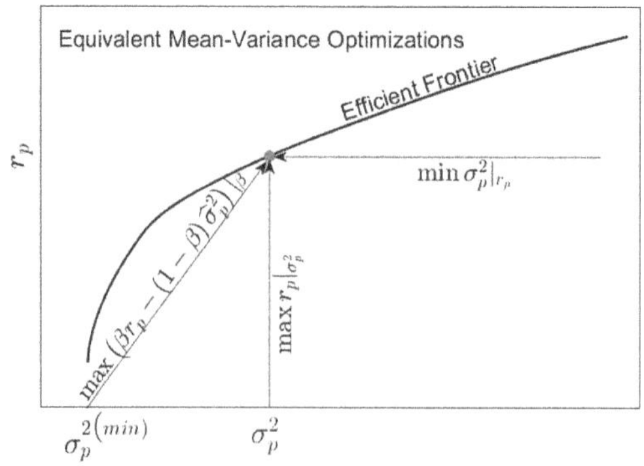

Fig. 3.2: Sketch indicating the three equivalent mean–variance optimization problems to produce a point on the efficient frontier

leads to the same system of linear equations as (3.3) with $q = 1/q_1$ and $\theta_0 = -\theta_1/q_1$. Writing the Lagrangian for (3.20) as

$$L(\boldsymbol{w}, \theta_2) = -\beta \overline{\boldsymbol{r}}^\top \boldsymbol{w} + (1-\beta)\boldsymbol{w}^\top \Sigma \boldsymbol{w} + \theta_2(1 - \boldsymbol{e}_n^\top \boldsymbol{w}),$$

produces the solution

$$\boldsymbol{w}^* = \frac{1}{2\beta}\Sigma^{-1}\overline{\boldsymbol{r}} + \frac{2\beta - C}{2\beta B}\Sigma^{-1}\boldsymbol{e}_n \tag{3.21}$$

where B and C are as defined in (3.5). Comparing solution (3.4) with (3.21), it is straightforward to show the equivalence

$$\overline{r}_p = \frac{\beta}{2(1-\beta)}\frac{\Delta}{B} + \frac{C}{B}. \tag{3.22}$$

The choice $\beta = 0$ gives $\overline{r}_p = \overline{r}_{p,\min} = C/B$ as derived in Section 3.1. In that section we noted that no solution can be obtained for the minimization (3.1) if the restriction $\overline{r}_p$ is set too large. Minimization (3.20) has the advantage that as $\beta \uparrow 1$ a solution is still obtained.[20] In this case the solution will be (in the absence of any other constraints) $w_i = \delta_{ik}$ where asset k has the largest average return value $\overline{r}_k$.

In Chapter 6, we will employ (3.20), using CVaR rather than variance as the risk measure of the portfolio.

3.2 CVaR–Minimizing Portfolios

We introduce mean–CVaR portfolio optimization, which replaces variance with CVaR as the risk measure of a portfolio. To understand the choice of CVaR as a risk measure, we first review the concept of coherent risk measures.

20 The feasible set for (3.20) is larger than that for (3.1).

3.2.1 Coherent Risk Measures

A coherent risk measure satisfies four essential properties:[21] monotonicity, sub–additivity, homogeneity, and translation invariance (Artzner et al. 1999).

1. *Monotonicity:* If portfolio A almost always (i.e., with probability 1) has higher returns than portfolio B, the risk associated with A must be lower.[22]
2. *Subadditivity:* The risk of a combined portfolio is less than or equal to the sum of the risks of the individual portfolios. Subadditivity implies a risk benefit under diversification.
3. *Homogeneity:* If the portfolio value is scaled by a positive factor, the portfolio risk scales by the same factor. (Colloquially, doubling the investment doubles the risk.)
4. *Translation invariance:* Adding a risk–free asset to a portfolio decreases the portfolio risk by (the risk equivalent of) the amount of capital held in the risk–free asset.

The portfolio standard deviation, which is used in mean–variance optimization, is not a coherent risk measure[23] as it does not satisfy monotonicity (Artzner et al. 1999). The risk measure value–at–risk (VaR)[24] fails to satisfy subadditivity (Artzner et al. 1999). The conditional value–at–risk, CVaR,[25] is a risk measure that is coherent. It provides a more accurate estimate of anticipated loss than does VaR, which provides only a lower bound on the loss amount.

While the standard deviation is a measure of the "central spread" (central risk) associated with a distribution, both VaR and CVaR are measures of "tail spread" (tail risk). Tail risk measures therefore have two specific advantages in financial application. They are measures that are sensitive to changes in tail heaviness and they concentrate specifically on what risk managers worry about most – the probability and size of loss. Consequently, their use leads to more informed decision making, better risk management practices, and more resilient investment strategies in markets characterized by heavy–tailed return distributions and extreme events (Rockafellar and Uryasev 2000; Inui and Kijima 2005).

The theoretical underpinnings and practical applications of CVaR are well documented. Mazaheri (2008) discusses risk budgeting using CVaR, highlighting its suitability for trades with heavy–tailed return distributions. Hardy (2006) introduces risk measures for actuarial applications, emphasizing the robustness of CVaR in financial risk

---

21 An additional normalization property requires that the risk measure be zero for any portfolio containing no risky assets.

22 Relative to B, portfolio A is acting like a riskless asset, generating higher returns with probability one.

23 The properties of coherent risk measures can also be extended to reward–risk ratios such as the Sharpe ratio (Cheridito and Kromer 2013).

24 VaR and CVaR are formally defined and illustrated in Section 3.2.2.

25 CVaR is also known variously as the expected shortfall (ES), average value–at–risk (AVaR), or expected tail loss (ETL).

management. Rachev et al. (2008) provide extensive insights into stochastic models and portfolio optimization, showcasing the application of CVaR in various financial contexts.

3.2.2 Mean–CVaR Portfolio Optimization

The mean–variance portfolio model discussed in Section 3.1 establishes a framework for developing similar strategies that use other measures as proxies for the portfolio risk. We describe portfolio optimization using CVaR as the chosen risk measure.[26] We begin with the formal definition of VaR. Let $F(x) = \Pr(\overline{r}_p \le x)$ denote the CDF of returns $\overline{r}_p$ of a portfolio. Then[27]

$$\mathrm{VaR}_\alpha(\overline{r}_p) = -\inf\{x \in \mathbb{R} | F(x) \ge \alpha\}\,. \tag{3.23}$$

The negative sign in (3.23) is a convention which reports VaR values for losses as positive. The value α is referred to as the tail probability. Conceptually, (3.23) says that: if the portfolio return $\overline{r}_p$ is a random variable with $\mathrm{VaR}_\alpha(\overline{r}_p) = x_\alpha$, then α percent of the time the portfolio return will be less than $-x_\alpha$; that is, the loss (expressed as a positive value) will exceed x_α.

$\mathrm{CVaR}_\alpha(\overline{r}_p)$ is defined as the average over all values $\mathrm{VaR}_\gamma(\overline{r}_p)$ where $0 \le \gamma \le \alpha$,

$$\mathrm{CVaR}_\alpha(\overline{r}_p) = \frac{1}{\alpha}\int_0^\alpha \mathrm{VaR}_\gamma(\overline{r}_p)\,d\gamma\,. \tag{3.24}$$

From (3.24) we have that, for any given value of α, $\mathrm{CVaR}_\alpha(\overline{r}_p) \ge \mathrm{VaR}_\alpha(\overline{r}_p)$.

As an illustration of VaR, an hourly value of $\mathrm{VaR}_{\alpha=0.05} = 40{,}000$ USD indicates there is a 5% probability that the hourly loss of a portfolio will exceed $40,000 (assuming there are no changes in the portfolio makeup over this period). Stated more positively, it indicates a 95% probability of either an hourly profit or of a loss that will be less than $40,000. In contrast, CVaR is a probability weighted average of all losses in the lowest 5% of the tail. Thus an hourly value of $\mathrm{CVaR}_{0.05} = 200{,}000$ USD indicates that, with 5% probability, the hourly loss of a portfolio is expected to be (exactly) $200,000.

The objective of CVaR portfolio optimization (Rockafellar and Uryasev 2000, 2002; Krokhmal et al. 2002; Tütüncü et al. 2003) is to achieve a target portfolio expected return having the smallest possible CVaR, with the CVaR most commonly measured at either the 95% or 99% confidence levels (i.e., $\alpha = 0.05$ or 0.01) in empirical application. The

---

26 The Basel III regulatory framework for banks mandates the use of $\mathrm{CVaR}_{\alpha=0.01}$ as the risk measure.
27 Definition (3.23) states the following. Given $\mathrm{VaR}_\alpha(\overline{r}_p) = x_\alpha$, then: $\inf\{x \in \mathbb{R}|F(x) \ge \alpha\} = -x_\alpha$; $F(-x_\alpha) = \alpha$; and the portfolio returns will be less than $-x_\alpha$ with probability α. The infinum in the definition (3.23) is required to uniquely define $\mathrm{VaR}_\alpha(x)$ in cases for which the CDF $F(x)$ (which is always a non–decreasing function) is either a càdlàg function or continuous, but constant, over a range of values of x.

CVaR_α–minimizing optimization problem is therefore

$$\min_{\boldsymbol{w}} \mathrm{CVaR}_\alpha(\boldsymbol{w}) \quad \text{subject to} \quad \begin{cases} \overline{\boldsymbol{r}}^\top \boldsymbol{w} = \overline{r}_p\,, \\ \boldsymbol{e}_n^\top \boldsymbol{w} = 1\,. \end{cases} \tag{3.25}$$

In contrast to the optimization problem (3.1), the objective function in (3.25) is not in quadratic form and must be manipulated to achieve a form convenient for optimization. We outline the original approach proposed by Rockafellar and Uryasev (2002) to achieve this. From the constraint $\overline{r}_p = \overline{\boldsymbol{r}}^\top \boldsymbol{w}$, we can interpret $\overline{r}_p$ as a random sample from the joint distribution function $f(\boldsymbol{w}, \boldsymbol{r})$. Let $p(\boldsymbol{r})$ denote the probability density function (PDF) that determines the asset returns $\boldsymbol{r}$. For any fixed $\boldsymbol{w}$, the CDF of the portfolio return is then

$$F(\boldsymbol{w}, x) = \int_{f(\boldsymbol{w},\boldsymbol{r}) \le x} p(\boldsymbol{r})\, \mathrm{d}\boldsymbol{r}. \tag{3.26}$$

From (3.26) and (3.23),

$$\mathrm{VaR}_\alpha(\boldsymbol{w}) = -\inf\{x \in \mathbb{R} | F(\boldsymbol{w}, x) \ge \alpha\}, \tag{3.27}$$

while from (3.24),

$$\mathrm{CVaR}_\alpha(\boldsymbol{w}) = \frac{1}{\alpha} \int_{-f(\boldsymbol{w},\boldsymbol{r}) \ge \mathrm{VaR}_\alpha(\boldsymbol{w})} [-f(\boldsymbol{w}, \boldsymbol{r})] p(\boldsymbol{r})\, \mathrm{d}\boldsymbol{r}. \tag{3.28}$$

The range of integration in (3.28) makes its implementation challenging. Combining (3.27) and (3.26) produces the relation

$$\alpha = F(\boldsymbol{w}, \mathrm{VaR}_\alpha(\boldsymbol{w})) = \int_{-f(\boldsymbol{w},\boldsymbol{r}) \ge \mathrm{VaR}_\alpha(\boldsymbol{w})} p(\boldsymbol{r})\, \mathrm{d}\boldsymbol{r}.$$

The following manipulations then produce a practical form of the utility function, involving an unrestricted integration range.

$$\begin{aligned} \mathrm{CVaR}_\alpha(\boldsymbol{w}) &= \mathrm{CVaR}_\alpha(\boldsymbol{w}) + \mathrm{VaR}_\alpha(\boldsymbol{w}) - \mathrm{VaR}_\alpha(\boldsymbol{w}) \\ &= \mathrm{VaR}_\alpha(\boldsymbol{w}) + \mathrm{CVaR}_\alpha(\boldsymbol{w}) - \frac{1}{\alpha} \cdot \alpha \mathrm{VaR}_\alpha(\boldsymbol{w}) \\ &= \mathrm{VaR}_\alpha(\boldsymbol{w}) + \frac{1}{\alpha} \int_{-f(\boldsymbol{w},\boldsymbol{r}) \ge \mathrm{VaR}_\alpha(\boldsymbol{w})} -f(\boldsymbol{w}, \boldsymbol{r}) p(\boldsymbol{r})\, \mathrm{d}\boldsymbol{r} \\ &\qquad - \frac{1}{\alpha} \mathrm{VaR}_\alpha(\boldsymbol{w}) \int_{-f(\boldsymbol{w},\boldsymbol{r}) \ge \mathrm{VaR}_\alpha(\boldsymbol{w})} p(\boldsymbol{r})\, \mathrm{d}\boldsymbol{r} \\ &= \mathrm{VaR}_\alpha(\boldsymbol{w}) + \frac{1}{\alpha} \int_{-f(\boldsymbol{w},\boldsymbol{r}) \ge \mathrm{VaR}_\alpha(\boldsymbol{w})} [-f(\boldsymbol{w}, \boldsymbol{r}) - \mathrm{VaR}_\alpha(\boldsymbol{w})]\, p(\boldsymbol{r})\, \mathrm{d}\boldsymbol{r} \\ &= \mathrm{VaR}_\alpha(\boldsymbol{w}) + \frac{1}{\alpha} \int [-f(\boldsymbol{w}, \boldsymbol{r}) - \mathrm{VaR}_\alpha(\boldsymbol{w})]^+\, p(\boldsymbol{r})\, \mathrm{d}\boldsymbol{r}, \end{aligned} \tag{3.29}$$

where $z^+ = \max(z, 0)$. If $f(\boldsymbol{w}, \boldsymbol{r})$ is a convex function[28] with respect to $\boldsymbol{w}$, then $\mathrm{CVaR}_\alpha(\boldsymbol{w})$ is also a convex function with respect to $\boldsymbol{w}$, indicating that there exists a global minimum value.

In practice, (3.29) is evaluated for a portfolio using the sample of asset returns $\boldsymbol{r}(t)$ over the set of discrete, historical times $t = \{t_1, t_2, \cdots, t_T\}$. The discrete form of (3.29) is

$$\mathrm{CVaR}_\alpha(\boldsymbol{w}) = \mathrm{VaR}_\alpha(\boldsymbol{w}) + \frac{1}{\alpha T} \sum_{t=t_1}^{t_T} \max\left(0, -\boldsymbol{r}(t)^\top \boldsymbol{w} - \mathrm{VaR}_\alpha(\boldsymbol{w})\right). \tag{3.30}$$

Using (3.30), the discrete form of the optimization problem (3.25) can be written

$$\min_{\boldsymbol{w},\boldsymbol{y}} \mathrm{VaR}_\alpha(\boldsymbol{w}) + \frac{1}{\alpha T} \sum_{t=t_1}^{t_T} y_t \quad \text{subject to} \begin{cases} y_t \geq \max\left(-\boldsymbol{r}(t)^\top \boldsymbol{w} - \mathrm{VaR}_\alpha(\boldsymbol{w}), 0\right), \\ \overline{\boldsymbol{r}}^\top \boldsymbol{w} = \overline{r}_p\,, \\ \boldsymbol{e}_n^\top \boldsymbol{w} = 1\,, \end{cases} \tag{3.31}$$

In (3.31), $\overline{\boldsymbol{r}} = T^{-1} \sum_{t=t_1}^{t_T} \boldsymbol{r}(t)$ is the vector of mean asset returns computed for the time period $[t_1, t_T]$ and $\boldsymbol{y} = (y_{t_1}, \ldots, y_{t_T})$. Note that the single constraint $y_t \geq \max\left(-\boldsymbol{r}(t)^\top \boldsymbol{w} - \mathrm{VaR}_\alpha(\right.$ can be written as two simpler linear constraints, $y_t \geq -\boldsymbol{r}(t)^\top \boldsymbol{w} - \mathrm{VaR}_\alpha(\boldsymbol{w})$ and $y_t \geq 0$.

Because $\mathrm{CVaR}_\alpha(\boldsymbol{w})$ is the average over all values $\mathrm{VaR}_\gamma(\boldsymbol{w})$ with $0 \leq \gamma \leq \alpha$, the value $\boldsymbol{w}$ that minimizes $\mathrm{CVaR}_\alpha(\boldsymbol{w})$ will also minimize $\mathrm{VaR}_\alpha(\boldsymbol{w})$. This leads to the following approach (Rockafellar and Uryasev 2000; Tütüncü et al. 2003):

$$\min_{\boldsymbol{w},\gamma} F_\alpha(\boldsymbol{w}, \gamma) \ \text{subject to}\ F_\alpha(\boldsymbol{w}, \gamma) = \gamma + \frac{1}{\alpha} \int \left(-f(\boldsymbol{w}, \boldsymbol{r}) - \gamma\right)^+ p(\boldsymbol{r})\, \mathrm{d}\boldsymbol{r}\,.$$

The optimization function $F_\alpha(\boldsymbol{w}, \gamma)$ is convex with respect to γ, and is convex with respect to $(\boldsymbol{w}, \gamma)$ if $f(\boldsymbol{w}, \boldsymbol{r})$ is convex with respect to $\boldsymbol{w}$ (Rockafellar and Uryasev 2000). Using this approach, (3.31) can be written

$$\min_{\boldsymbol{w},\boldsymbol{y},\gamma} \gamma + \frac{1}{\alpha T} \sum_{t=t_1}^{t_T} y_t \quad \text{subject to} \begin{cases} y_t \geq \max\left(-\boldsymbol{r}(t)^\top \boldsymbol{w} - \gamma, 0\right), \\ \overline{\boldsymbol{r}}^\top \boldsymbol{w} = \overline{r}_p\,, \\ \boldsymbol{e}_n^\top \boldsymbol{w} = 1\,. \end{cases} \tag{3.32}$$

In (3.32), both the objective function and the constraints are linear in $\boldsymbol{w}$ and γ, resulting in a linear programming problem which can be solved using elementary methods, such as the simplex method.

As noted by Tütüncü et al. (2003), the constraint on y_t ensures only that $y_t \geq \max\left(0, -\boldsymbol{r}(t)^\top \boldsymbol{w} - \mathrm{VaR}_\alpha(\boldsymbol{w})\right)$. However, the minimization of the objective function ensures that, at the minimizing value $\boldsymbol{w}^*$,

$$y_t = \max\left(0, -\boldsymbol{r}(t)^\top \boldsymbol{w}^* - \mathrm{VaR}_\alpha(\boldsymbol{w}^*)\right).$$

28 A function $f(x)$ with domain I is convex if $f(\lambda x_1 + (1-\lambda)x_2) \leq \lambda f(x_1) + (1-\lambda) f(x_2)$, for any $x_1, x_2 \in I$ and $\lambda \in [0, 1]$.

As the parameter $\bar{r}_p$ varies, the optimizing solution $(\mathrm{CVaR}_\alpha(\boldsymbol{w}^*), \bar{r}_p(\boldsymbol{w}^*))$ produces a portfolio frontier curve in (risk, return) space analogous to that under mean–variance optimization. The portfolio frontier will have convex form, with a global minimum–risk portfolio having the value $\mathrm{CVaR}_{\alpha,\min}$ such that $\mathrm{CVaR}_\alpha(\boldsymbol{w}^*) \geq \mathrm{CVaR}_{\alpha,\min}$ for every minimizing solution $\boldsymbol{w}^*$. The portfolio $\boldsymbol{w}^*_{\min}$ satisfying $\mathrm{CVaR}_\alpha(\boldsymbol{w}^*) = \mathrm{CVaR}_{\alpha,\min}$ is known as the *minimum* CVaR_α *portfolio*. All the points on the portfolio frontier having $\bar{r}_p(\boldsymbol{w}^*) \geq \bar{r}_p(\boldsymbol{w}^*_{\min})$ comprise the efficient frontier. There is also a risk–maximizing portfolio defining the upper end of the mean–CVaR efficient frontier. Again, the MATLAB function `estimateFrontierLimits()` can be used to determine both end points for a CVaR minimizing optimization subject to a given set of constraints.

3.2.3 Capital Market Line and the Mean–CVaR Tangent Portfolio

We again consider the effect of adding a single riskless asset having guaranteed return r_f to a portfolio of n risky assets. The CVaR–minimizing optimization problem is

$$\min_{\boldsymbol{w}} \ \mathrm{CVaR}_\alpha(\boldsymbol{w}) \quad \text{subject to} \ (1 - \boldsymbol{e}_n^\top \boldsymbol{w})r_f + \bar{\boldsymbol{r}}^\top \boldsymbol{w} = \bar{r}_p \ . \tag{3.33}$$

Using the result (3.32), (3.33) can be expressed as

$$\min_{\boldsymbol{w},y,\gamma} \ \gamma + \frac{1}{\alpha T}\sum_{t=t_1}^{t_T} y_t \quad \text{subject to} \begin{cases} y_t \geq \max\left(-\boldsymbol{w}^\top \boldsymbol{r}(t) - (1 - \boldsymbol{e}_n^\top \boldsymbol{w})r_f - \gamma, 0\right), \\ (1 - \boldsymbol{e}_n^\top \boldsymbol{w})r_f + \bar{\boldsymbol{r}}^\top \boldsymbol{w} = \bar{r}_p \ . \end{cases}$$

At the minimizing values $\boldsymbol{w}^*$, $\gamma^* = \mathrm{VaR}_\alpha(\boldsymbol{w}^*)$, either $y_t = 0$ or $y_t = -(\boldsymbol{w}^*)^\top \boldsymbol{r}(t) - (1 - \boldsymbol{e}_n^\top \boldsymbol{w}^*)r_f - \gamma^* > 0$. If we define $S = \{t \mid y_t > 0\}$, then

$$\mathrm{CVaR}_\alpha(\boldsymbol{w}^*) = \gamma^* + \frac{1}{\alpha T}\sum_{t\in S}\left[-r_{p,t}(\boldsymbol{w}^* - \gamma^*)\right] = \left[1 - \frac{\mathrm{Card}(S)}{\alpha T}\right] - \frac{f_s}{\alpha}\bar{r}_p, \tag{3.34}$$

where $f_s = (T\bar{r}_p)^{-1}\sum_{t\in S} r_{p,t}(\boldsymbol{w}^*)$ and $\mathrm{Card}(S)$ denotes the number of elements in S. Because $\bar{r}_p = T^{-1}\sum_{t=t_1}^{t_T} r_{p,t}(\boldsymbol{w}^*)$, we see that f_s is just the fraction of $\bar{r}_p$ contributed by the returns in the set S.

From (3.34), we see that the trajectory of $(\mathrm{CVaR}_\alpha(\boldsymbol{w}^*), \bar{r}_p)$ is a straight line – the CML in CVaR–return coordinates. As $\bar{r}_p - r_f = (\bar{\boldsymbol{r}} - r_f\boldsymbol{e}_n)^\top \boldsymbol{w}^*$, where $(\bar{\boldsymbol{r}} - r_f\boldsymbol{e}_n)$ is a constant vector, $\boldsymbol{w}^*$ scales linearly with $\bar{r}_p - r_f$. If $\boldsymbol{w}^* = 0$, there are no risky assets in the portfolio and the risk $\mathrm{CVaR}_\alpha(\boldsymbol{w}^*) = 0$. The slope of the straight line is the stable tail adjusted return (STAR) ratio defined in (4.11). This line touches the efficient frontier tangentially at the point $\left(\mathrm{CVaR}_\alpha(\boldsymbol{w}^*_s), \bar{\boldsymbol{r}}^\top \boldsymbol{w}^*_s\right)$, where $\boldsymbol{w}^*_s$ satisfies $\boldsymbol{e}_n^\top \boldsymbol{w}^*_s = 1$. The portfolio having weights $\boldsymbol{w}^*_s$ is known as the *tangent* CVaR_α portfolio. Of all the portfolios on the efficient frontier, the tangent portfolio has the largest STAR ratio.

It is useful to express the CVaR and VaR in terms of the $100(1-\alpha)\%$ confidence level. For example, a tail probability (significance level) of $\alpha = 0.05$ corresponds to a 95%

confidence level. Thus, $\text{VaR}_{0.05}(r)$, referring to a tail probability of 5%, is often written $\text{VaR}_{95}(r)$. The notation $\text{VaR}_{0.05}(\overline{r}_p)$ implies a 5% probability that returns will be worse than $x_{0.05}$, while the notation $\text{VaR}_{95}(\overline{r}_p)$ conveys a more positive statement that there is a 95% probability that returns will be better than $x_{0.05}$. We will use both conventions (i.e., CVaR_{95} and $\text{CVaR}_{0.05}$) interchangeably. Which is in use should be readily apparent from context.

3.2.4 Empirical Application of Mean–CVaR Optimization using MATLAB

We demonstrate the application of CVaR–minimizing optimization to the same portfolio of cryptocurrencies considered in Section 3.1.2. We again illustrate using MATLAB code. As in Section 3.1.2, the purpose is to compute mean–CVaR optimized portfolio weights to be used using the time period 05:00:00 through 06:00:00 on September 8, 2023.

The first code block consists of Listing 3.1, which loads the price data and computes arithmetic returns. The second code block consists of Listing 3.3 to identify the risk–free rate appropriate for the time 05:00:00 on September 8, 2023. The third code block, presented below, establishes the CVaR portfolio object[29] for the optimization (3.32) subject to a long–only strategy.

Listing 3.7: Define CVaR portfolio object

```
% Create a CVaR portfolio object. Input the asset historical returns,
% set the risk-free rate, and record the asset names.
p = PortfolioCVaR('Scenarios', ret, 'RiskFreeRate', rf_rate);
p = p.setAssetList(cr_nms);

% Set default constraints. These constrain the weights to be
% non-negative and to sum to unity.
p = setDefaultConstraints(p);

% Set the desired confidence level.
alpha = 0.95;
p = setProbabilityLevel(p, alpha);

% Set the CVaR optimization solver. (This is the default solver.)
pf = setSolver(pf,'TrustRegionCP');
```

The fourth code block computes the efficient frontier, the CML, and the tangent portfolio.

29 There are slight differences between the MATLAB objects for mean–variance and CVaR portfolios. In particular, the CVaR portfolio structure requires assignment of the probability level, measured as $1 - \alpha$. See https://www.mathworks.com/help/finance/portfoliocvar.html for more details.

Listing 3.8: Compute CVaR efficient frontier

```
% Estimate portfolio weights on the efficient frontier using 50
% portfolios
portfolioWeights = estimateFrontier(p, 50);

% Compute [CVaR_p, r_p] for the efficient frontier portfolios.
% [EF_Risks(1,1), EF_Rtns(1,1)] is the global minimum portfolio.
% Note that the routine estimatePortMoments() used in Listing~\ref{lst:
    mv_5}
% is specific to mean-variance portfolios.
EF_Rtns = estimatePortReturn(p, portfolioWeights);
EF_Risks = estimatePortRisk(p, portfolioWeights);

% Find the portfolio with the highest STAR ratio.
% Sample the 50 portfolios as, unlike the mean-variance case,
% MATLAB does not provide a function to do this for mean_CVaR.
Sort_ratio = (EF_Rtns - rf_rate) ./ EF_Risks;
[maxSort, idxSort] = max(Sort_ratio);
Tan_Risk = EF_Risks(idxSort);
Tan_Rtn = EF_Rtns(idxSort);
```

The final code block plots the CVaR efficient frontier, CML, tangent portfolio and individual cryptocurrencies represented by their values $(\text{CVaR}_{95,i}, \overline{r}_i)$. Modifications to Listing 3.6 are required in order to plot the individual cryptocurrencies. Listing 3.9 provides the full code block for the plot.

Listing 3.9: Plot CVaR Efficient Frontier

```
pfig = figure();

% Plot the efficient frontier as a black solid line
plot(EF_Risks, EF_Rtns, 'k-', 'LineWidth', 2);
hold on;

% Plot the capital market line as a red dashed line.
x = [0 Tan_Risk];
y = [rf_rate Tan_Rtn];
plot(x, y, 'r--', 'LineWidth', 2);
hold on;

% Identify the tangent portfolio.
plot(Tan_Risk, Tan_Rtn, 'ro', 'Markersize', 8);
hold on;
```

```
% Empirically compute CVaR_95 for each asset
N_cr = length(cr_nms);
cvar_values = zeros(1, N_cr);

for i = 1:N_cr
    sorted_returns = sort(ret(:, i));
    index = ceil((1 - alpha) * length(sorted_returns));
    cvar_values(i) = -mean(sorted_returns(1:index));
end

% plot individual assets
mu = mu'; % Convert back to column vector.
scatter(cvar_values, mu, 'filled');

% Specify plot screen position and length/width (in points).
set(pfig, 'Position', [50, 250, 750, 500]);

% Specify default fontsize, axes labels and legend
fontsize(pfig, 14, 'points');
xlabel('$\textrm{CVaR}_{95}$', 'Interpreter', 'latex', 'FontSize',14);
ylabel('Expected Return', 'FontSize', 14);
legend({'Eff. Front.', 'CML', 'Tangent'}, 'Location', 'northwest');

% Add grid lines.
grid on;
```

The resultant plot is displayed in Fig. 3.3. Efficient frontier portfolios vary from the minimum risk portfolio $(0.015, 7.5 \cdot 10^{-6})$ to the maximum endpoint $(0.029, 9.7 \cdot 10^{-5})$. The tangent portfolio $(0.026, 9.0 \cdot 10^{-5})$ has an hourly STAR ratio of $3.2 \cdot 10^{-3}$. The cryptocurrency (SHIB) corresponding to the maximum limit of the efficient frontier is indicated by an arrow.

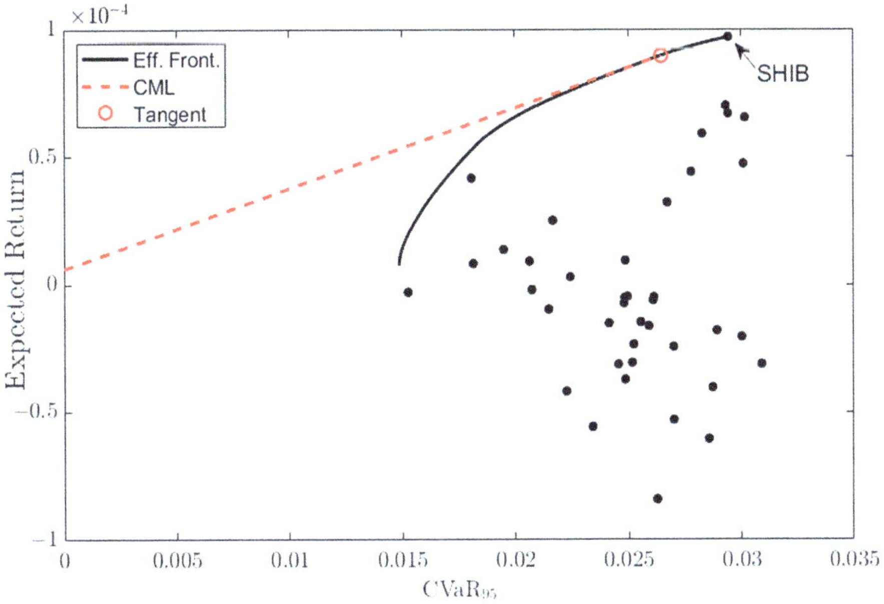

Fig. 3.3: Cryptocurrency mean–CVaR portfolio efficient frontier at 05:00:00, September 8, 2023 (Eff. Front.: efficient frontier; CML: capital market line; Tangent: tangent portfolio. Filled circles correspond to historical average return and $CVaR_{95}$ of individual crypto assets)

4 Historical Portfolio Optimization

This chapter presents a collection of optimized cryptocurrency–based portfolios, which serve as representative models of strategies employed by institutional investment managers in actively managed portfolios. These prototype portfolios offer diverse risk–return profiles, making them valuable tools for asset allocation across various market conditions. As appropriate, we compare the performance of the six portfolio optimizations derived in Sections 3.1 and 3.2 to the EWP and EWBH benchmarks derived in Section 2.1.1. For brevity, the six optimized portfolios are referred to as:

MVP minimum mean–variance portfolio (Section 3.1);
TVP tangent mean–variance portfolio (Section 3.1);
M95 minimum CVaR_{95} portfolio (Section 3.2);
T95 tangent CVaR_{95} portfolio (Section 3.2);
M99 minimum CVaR_{99} portfolio (Section 3.2); and
T99 tangent CVaR_{99} portfolio (Section 3.2).

CVaR optimizations at both the 95% and 99% quantile levels are included to analyze the impact of tail risk information on portfolio performance. For instance, in a 2,000–trading–hour sample of returns for a crypto asset, the CVaR_{99} tail–risk value is computed from only 20 return values, while CVaR_{95} uses 100 return values. This difference can lead to significant variations in portfolio behavior when optimized under CVaR_{95} versus CVaR_{99}.

This chapter focuses on portfolios composed of the cryptocurrencies introduced in Chapter 1. Optimal portfolio weights are determined using the historical hourly return data assuming constant volatility within each historical window. This assumption will be relaxed in Chapter 5. Optimization constraints are applied to model asset allocation controls and transaction costs.

Due to the computational intensity of dynamic optimization (Chapter 5), whose performance will be compared to historical optimization, we limit our analysis to the first 3,001 hourly data points (from 06:00:00 on July 29, 2021 to 13:00:00 on December 1, 2021) of cryptocurrency prices from Chapter 1. This reduces to 3,000 hourly returns. For optimizations, we employ a moving window of 2,000 hours of data returns,[1] which results in 1,000 weight optimizations covering the period 22:00:00 on October 20, 2021 to 13:00:00 on December 1, 2021. This time period was chosen as it corresponds to the November, 2021 peak and initial decline in cryptocurrency prices discussed in Chapter 1.

This chapter is organized as follows. In Section 4.1, we evaluate the performance of the six portfolio optimizations, as well as the EWP and EWBH benchmarks, in terms of cumulative return under four investment strategies. Except for the momentum strat-

1 The moving window must be sufficiently large to provide adequate statistical assessment of the set of historical return values.

https://doi.org/10.1515/9781501517136-004

egy, portfolio rebalancing is performed hourly. Given the high transaction costs associated with frequent rebalancing, Section 4.2 introduces turnover as a cost proxy and examines performance under increasing turnover constraints. Section 4.3 introduces risk measures designed to quantify the relative risks of different portfolio–strategy combinations.

4.1 Basic Strategies: Return and Weight Performance

In addition to the long–only investment strategy (Section 4.1.1), we evaluate the performance of the portfolio optimizations under a general long–short strategy (Section 4.1.2) and a restricted 130/30–type strategy (Section 4.1.3). In Section 4.1.4, we explore the implications of a momentum strategy, where rebalancing occurs less frequently.

The weights for individual cryptocurrencies in each portfolio are determined based on returns calculated from a rolling window of 2,000 trading hours. (For instance, the optimized weights used for portfolio return computations at 22:00:00 on October 20, 2021 are derived from data spanning the previous 2,000 trading hours, specifically from 7:00:00 on July 29, 2021 to 21:00:00 on October 20, 2021). By iterating this recursive process, an hourly sequence of optimized portfolio weights is generated covering the period from 22:00:00 on October 20, 2021 to 13:00:00 on December 1, 2021. Once the complete time series of optimal portfolio weights is constructed, performance measures are calculated. It is important to recall (Section 2.1.1) that the methodology for the EWP and EWBH benchmarks is significantly simpler. These strategies do not involve a 2,000-hour rolling window or weight optimization.

Prior to discussing strategies, we consider the question of the choice of the initial (t_0) asset distribution (i.e. the asset weights $w_i(t_0)$) of a portfolio. There is no definitive answer as the question must concern itself with either an existing portfolio composition or a newly composed portfolio.

We consider the problem of starting a new portfolio at time t_0. Specifically the first period of trading of the assets in the new portfolio covers the interval $(t_0, t_1]$, with initial asset weights $w_i(t_0)$, $i = 1, \ldots, n$, representing a selection of n assets to be purchased (or shorted) at t_0. To obtain $w_i(t_0)$, $i = 1, \ldots, n$, the optimization algorithm must be run $p \geq 1$ times, using historical data over a time period $[t_0 - \tau + 1, t_0]$ to generate a sequence of weights $w_i(k)$, $k = 1, \ldots, p$, $i = i, \ldots, n$. The value of p will depend on the constraints. This sequence is computed with the weights $w_i(k-1)$ used as input for the optimization producing the next set $w_i(k)$, until all traces of the initial starting value $w_i(k = 1)$ (i.e. of the initial transient) are removed. Then $w_i(t_0) = w_i(p)$ serves as the optimized starting portfolio. Only once the weights $w_i(t_0)$ are known, are assets purchased or short contracts set up to establish the new portfolio.

The more difficult problem consists of starting with a predetermined portfolio. As an "inherited" portfolio is virtually never optimized, there will be a transient period – dependent on constraints – over which the portfolio approaches true optimization.

Under the long–only constraints of Section 4.1.1, the optimizer can produce a fully optimized set of weights after the first time interval. However, under: highly constrained long–short strategies (Jacobs et al., Section 4.1.2 or Lo–Patel, Section 4.1.3); momentum strategies (Section 4.1.4); or turnover constraints (Section 4.2), the transient period can become very lengthy. In such cases the benefits of the strategy may be wasted, with the strategy generating excessive transaction costs in order to transform the non–optimized initial portfolio into a strategy–optimized one.

In Sections 4.1.1–4.1.3 we consider starting with an inherited portfolio. As there are too many possibilities for the composition of such a portfolio, we assume the case of an equal–weighted portfolio. In Section 4.1.4, to clearly isolate momentum effects, we consider the problem of starting with a new portfolio. The transient sequence $w_i(k)$, $k = 1, \ldots, p$, used to generate $w_i(t_0)$ started with $w_i(k = 1) = 1/n$, $i = 1, \ldots, n$.

4.1.1 Long–Only Strategy

In this section, we examine the basic long–only strategy, where all investments are allocated to assets and asset weights are constrained to the range $[0, 1]$,

$$\sum_{i=1}^{n} w_i(t) = 1, \quad 0 \leq w_i(t) \leq 1. \tag{4.1}$$

For now we assume that transaction costs are negligible. The cumulative arithmetic return, from 22:00:00 on October 20, 2021 to 13:00:00 on December 1, 2021, of each optimized portfolio and the two benchmark portfolios is presented in Fig. 4.1. It is necessary to be conversant with the cumulative return (as opposed to the price) performance of a portfolio (or indeed, any asset). The cumulative return designates how well an investment performed independent of the amount of the initial investment. For example, an initial investment at 22:00:00 on October 20, 2021 in the T99 portfolio led to a cumulative return of 1.23 by 13:00:00 on December 1, 2021. From (2.10), this indicates that an initial investment of P_0 in this portfolio has resulted in a final wealth of 2.23 P_0, a 123% gain in wealth.

The three tangent portfolios, TVP, T95, and T99, consistently outperform all other portfolios. Notably, T99 consistently surpasses both TVP and T95 in performance. In contrast, the three risk–minimized portfolios, MVP, M95, and M99, exhibit closely aligned performance trajectories (they appear as a single trajectory in the figure). They consistently underperform both of the EWP and the EWBH strategies. Interestingly, prior to November 16, 2021, the performance of the non–optimized EWP and EWBH strategies showed little difference. However, after November 16, 2021, the EWBH strategy began to outperform the EWP.

Fig. 4.2 shows a stacked bar graph representation of the asset weights as a function of time for the MVP, M99, TVP and T99 portfolios. (These are chosen to represent central (variance) risk and extreme tail risk ($CVaR_{99}$). The optimized M99 portfolio consists of a

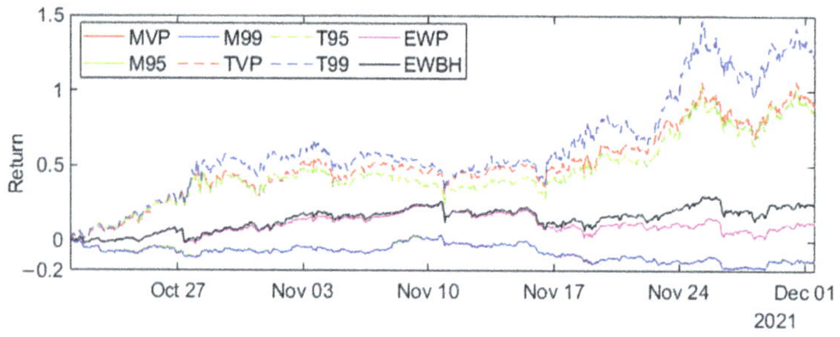

Fig. 4.1: Cumulative arithmetic return for the long–only strategy compared to those for the benchmarks

single crypto asset (BTC) (except for one time point, the 02:00:00 November 30, 2021 optimization, which included SOL at a weighting of 0.008). Thus the cumulative return series for M99 shown in Fig. 4.1 also holds for the price series for BTC in Fig. 1.1. The optimized MVP portfolio consisted largely of BTC with a small admixture of BNB, QNT, TRX and XMR. In contrast to the minimum–risk portfolios, which weight BTC very heavily, the tangent portfolios concentrate their optimized weights relatively equally among SOL, SHIB and SAND. Smaller additions consist of AVAX and AXS (prior to November 11, 2021) and AVAX and EGLD (after November 17, 2021). The crypto market began its downward trend in the period in November, 2021. (SHIB prices peaked on October 28, BTC and SOL on November 7, SAND on November 25.) Under long–only optimization, BTC is "seen" as the minimum–risk choice under both variance and extreme tail risk while SAND, SHIB and SOL are the corresponding minimizers under the tangent portfolio optimization.

No trace of the initial equal–weighted portfolio is evident in these weight plots; the long–only optimization is capable of generating a fully optimized weight solution each time it is called.

Fig. 4.2 also provides plots of each asset's average weight over the time period indicated in Fig. 4.2. As few assets have non–zero weightings,[2] the plot merely serves to emphasize the differences noted between the minimum–risk and tangent portfolios.

For comparison, Fig. 4.3 provides the weight time series and asset average weights for the EWBH portfolio. Only seven assets (AVAX, BAT, EGLD, ENJ, SAND, SHIB, and SOL) have a time–averaged weight above the equal–weighted value of 0.025, with SAND being dominant. Five assets (BNB, GRT, MATIC, NEAR, VET) maintain time averages essentially at the equal–weighted value. The remaining assets, including BTC, lie below the equal–weighted value.

2 Due to round off errors, values of $o(10^{-10})$ are consistent with zero.

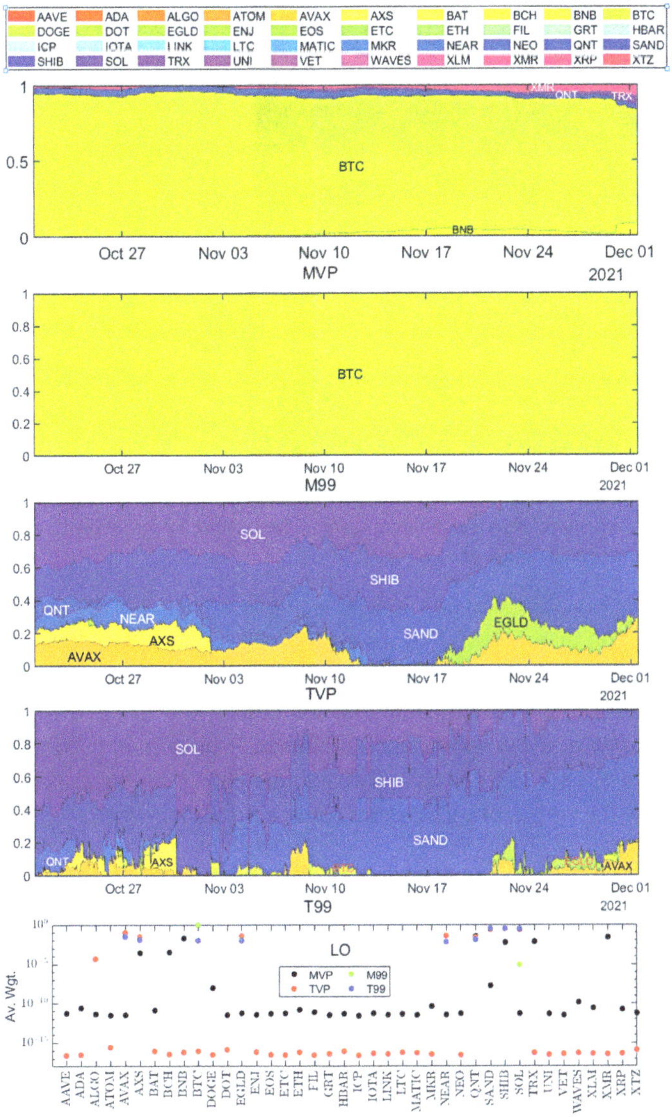

Fig. 4.2: Asset weights as a function of time for the MVP, TVP, M99 and T99 optimizations subject to the long-only strategy with no turnover constraint. (bottom) Time-averaged asset weights

4.1.2 Jacobs et al. Long-Short Strategy

Intuitively, a long–short optimized strategy has the potential to enhance the expected return of a portfolio for a given level of volatility. (However, short selling introduces substantial additional risk, including the potential for: heavy losses due to price increase of the shorted assets, unexpected changes in borrowing fees, loss of borrowed

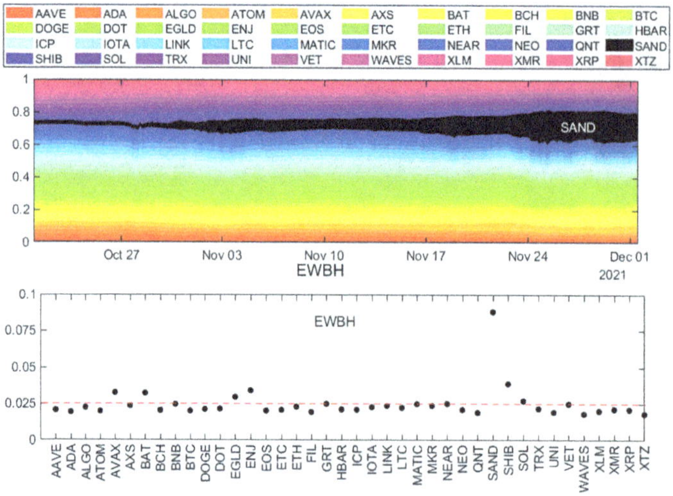

Fig. 4.3: (top) Asset weights as a function of time and (bottom) average asset weight for the EWBH portfolio (The red dashed line indicates the equi-weighted value of 0.025)

asset dividend payments, and margin calls.) Jacobs et al. (1999) provide a comprehensive overview of long–short portfolio management and adapt the MVP model to accommodate long–short portfolios, where the optimization of long and short positions is conducted simultaneously (separate optimization results in a suboptimal portfolio). Consequently, we consider the six MPT-based portfolios[3] under a Jacobs et al. (1999) long–short strategy, where the weight of each asset is constrained as follows:

$$\sum_{i=1}^{n} w_i(t) = 1, \quad -s \leq w_i(t) \leq 1 + s. \tag{4.2}$$

Equation (4.2) allows any asset to be shorted up to $s \cdot 100\%$ of the total portfolio weight or held long up to $(1 + s) \cdot 100\%$. As asset weights are recalculated hourly, long and short positions are rebalanced hourly without accounting for transaction costs, margin-related expenses (such as maintenance margins or interest payments), or the availability of assets (e.g., through brokers).

Fig. 4.4 presents the cumulative return of the portfolios under an aggressive parameter setting of $s = 0.3$. This long–short strategy underperformed compared to the long-only strategy (Fig. 4.1). There is a notable difference in the behavior of the T99 and TVP tangent portfolios under the two strategies. At the conclusion of the study period, the TVP portfolio yielded the highest cumulative return (85%) under the long–short strategy. This contrasts with the long-only strategy, where the T99 portfolio achieved the highest cumulative return (130%). Additionally, the tangent portfolios in the long–short strategy

3 Since the EWP and EWBH portfolios are long-only, they are excluded from this comparison.

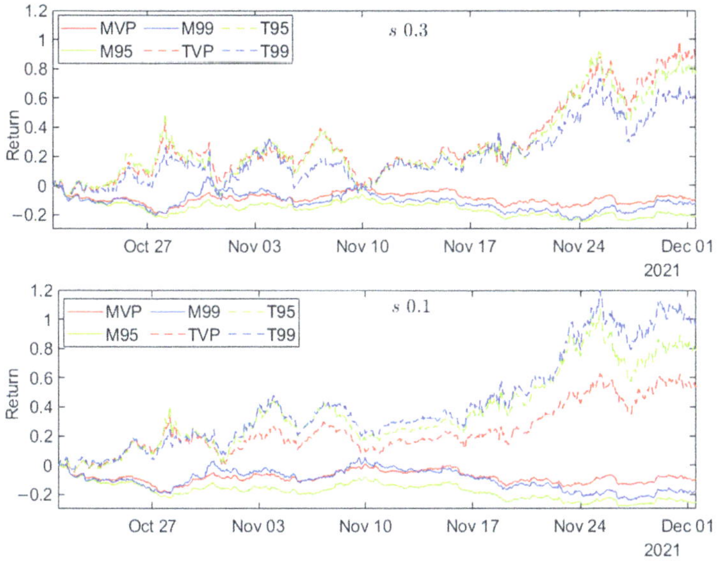

Fig. 4.4: Cumulative return for the Jacobs et al. strategy constrained by (4.2) with (top) $s = 0.3$ and (bottom) $s = 0.1$

exhibited greater cumulative return volatility than their counterparts in the long–only strategy. While there was no significant difference in the behavior of the minimum–risk portfolios under the long–only strategy, clear separation of the performance of these portfolios appeared under the long–short strategy.

The suboptimal performance of the Jacobs et al. methodology underscores the necessity for meticulous execution in long–short strategies. Our prototype portfolio, comprising 40 assets, is relatively modest in scale. The decision to permit short selling of up to 30% for one or more assets represents a significant risk exposure. Fig. 4.4 also illustrates the cumulative return of optimized portfolios when the parameter s in (4.2) is adjusted to 0.1. There is a flip in the order of cumulative return of the tangent portfolios compared to $s = 0.3$. Under the revised parameter value, the T99 portfolio emerged as the top performer, with a 100% cumulative return, while TVP showed a degraded cumulative return of 54%. The behaviors of the minimum–risk portfolios M95 and M99 were further degraded under the reduction in the value of s, while the cumulative return of MVP remained relatively unchanged.

Fig. 4.5 shows the asset weights as a function of time for the MVP, M99, TVP and T99 portfolios. Only the dominant, positive weighted assets are identified. The results echo those for the long–only strategy in the sense that BTC has a significant positive weight for the minimum–risk portfolios, while SOL, SHIB and SAND (along with BTC) has significant positive weights in the tangent portfolios. Unlike the long–only case, allowing for asset shorting results in every asset having a non–negligible weight (either positive or negative). Fig. 4.6 plots each asset's average weight over the time period indicated in

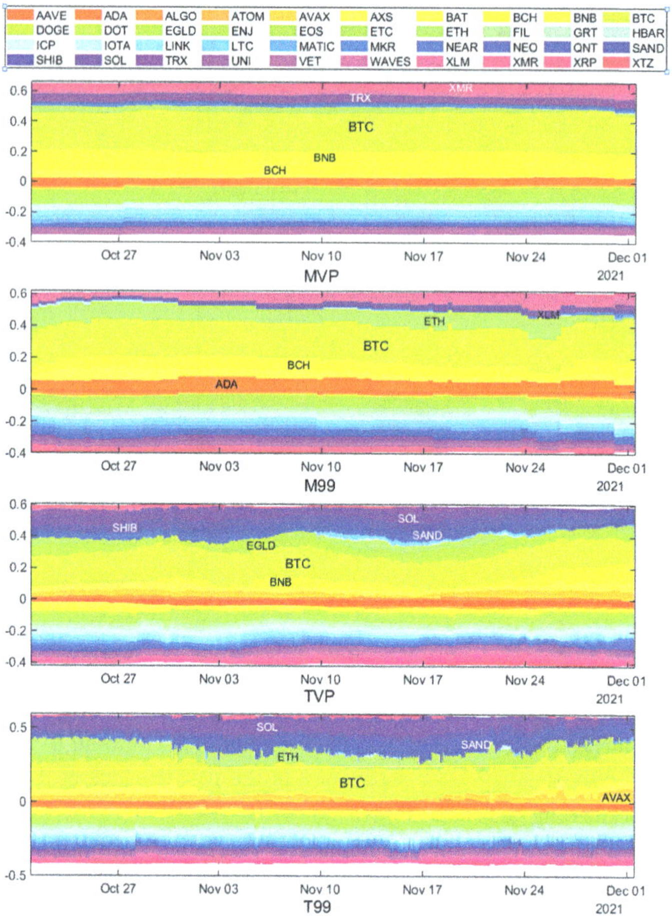

Fig. 4.5: Asset weights as a function of time for the MVP, TVP, M99 and T99 optimizations under the Jacob et al. strategy with $s = 0.1$

Fig. 4.5. While there are differences in average weightings among the four portfolios, they are much less pronounced than in Fig. 4.2 for the long–only strategy. No trace of the initial equal–weighted portfolio is evident in these weight plots.

Not so apparent in the average asset weight plot of Fig. 4.5 is the fact that, over time, assets may be alternately shorted or kept in a long position in the portfolio. Fig. 4.6 also plots the fraction of the hourly time periods that each asset is shorted. (A value of one indicates an asset is always shorted, zero indicates never shorted.) There are varied differences between the four optimizations and the assets. For examples: for all four optimizations BTC is never shorted, while AAVE, EOS, GRT, ICP, and VET are always shorted; FIL, MKR and XMR are never shorted under MVP, and always shorted under T99.

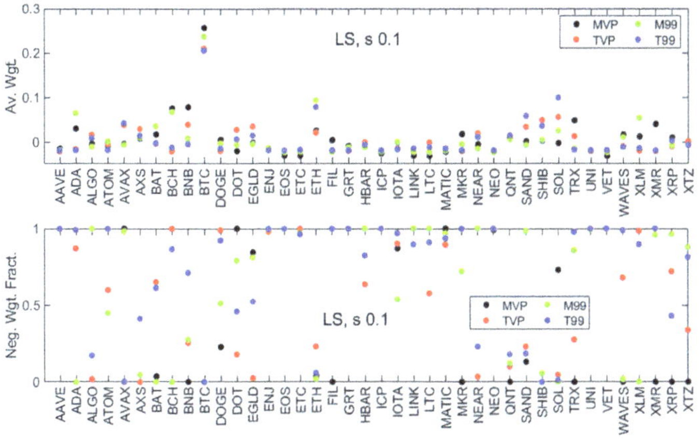

Fig. 4.6: (top) Average asset weight and (bottom) fraction of hourly time period an asset is shorted over the time period 22:00:00 on October 20, 2021 to 13:00:00 on December 1, 2021 for the indicated optimizations under the Jacobs et al. strategy with $s = 0.1$

4.1.3 Lo-Patel Long-Short Strategy

We next examine a 130/30–inspired long–short strategy, where 30% of the initial capital is obtained through shorting, and 130% of the starting capital is allocated to long positions. Lo and Patel (2008) constructed such a 130/30 equity portfolio using the S&P 500 universe of stocks and a "standard"[4] portfolio optimizer. Their strategy, which involves monthly returns and rebalancing, generates a benchmark time series of returns that they interpret as a 130/30 index. The 130/30 strategy is a leveraged strategy defined by the following constraints:

$$\sum_{i=1}^{n} w_i(t) = 1, \qquad lev \le w_i(t) \le 1 + lev, \tag{4.3a}$$

$$\sum_{i=1}^{n} \max\left(0, w_i(t) - w_i(t-1)\right) \le 1 + lev, \tag{4.3b}$$

$$\sum_{i=1}^{n} \max\left(0, w_i(t-1) - w_i(t)\right) \le lev, \tag{4.3c}$$

where $lev = 0.3$. This strategy differs from the one proposed by Jacobs et al. in that it imposes restrictions on both the sum of positive weight changes (4.3b) and the sum of negative weight changes (4.3c). Guerard et al. (2010), among others, have evaluated the performance of 130/30 models against long-only models using mean–variance optimization.

4 Specifically, the MSCI Barra Aegis Portfolio Manager with the Barra Equity Long–Term Risk Model.

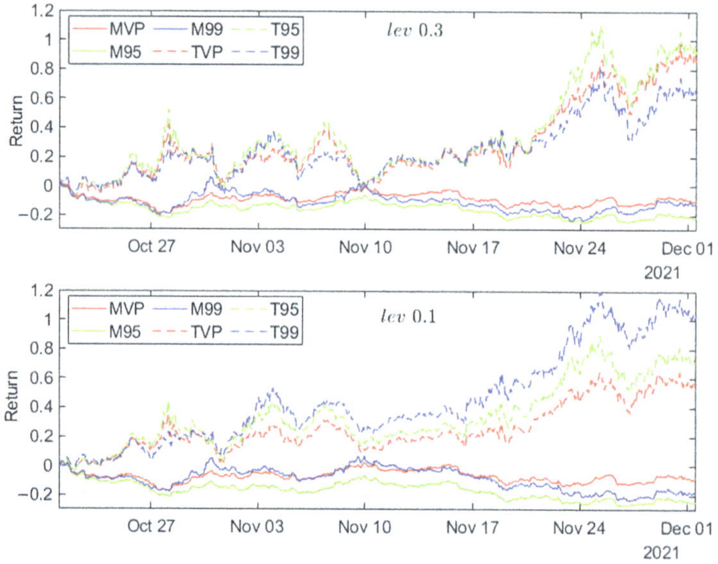

Fig. 4.7: Cumulative return for the Lo–Patel strategy with (top) $lev = 0.3$ and (bottom) $lev = 0.1$

Fig. 4.7 compares the cumulative return plots for optimizations conducted under leverage values of $lev = 0.3$ and $lev = 0.1$. The performance of the portfolios under the two leverage values is similar, the major difference being in the relative performance of TVP/T95 versus T99. The volatility and drawdown (see (4.9)) of the tangent portfolios are higher under $lev = 0.3$.

Fig. 4.8 provides the asset weights as a function of time for the MVP, M99, TVP and T99 portfolios under this Lo–Patel strategy. After an initial time period of roughly 24 hours, the asset weights for the Jacobs et al. strategy with $s = 0.1$ and the Lo–Patel strategy with $lev = 0.1$ become very similar. This similarity is seen by comparing the asset average weights and negative weight fractions in Fig. 4.9 for the Lo–Patel strategy with the corresponding plots in Fig. 4.6 for the Jacob's et al. There are minor differences. strategy. Under the Jacob's et al. strategy: for all four optimizations only BTC is never shorted, while only ICP is always shorted; a single asset, MKR, is never shorted under MVP, and always shorted under T99.

The transient evident in the Lo–Patel weight plots with $lev = 0.1$ results from the fact that the restriction (4.3) on weight changes is much stricter than that of Jacobs et al. (4.2), which is on weight values. The Lo–Patel constraints are a form of turnover constraint (Section 4.2). Thus a sequence of roughly $p = 24$ initial optimizations under a Lo–Patel $lev = 0.1$ strategy would be required to transform an initial equal–weighted selection of n assets into a fully optimized initial set of weights. It is therefore interesting that, post transient, the weight distributions under the Jacobs et al. strategy with $s = 0.1$ and Lo–Patel with $lev = 0.1$ are very similar.

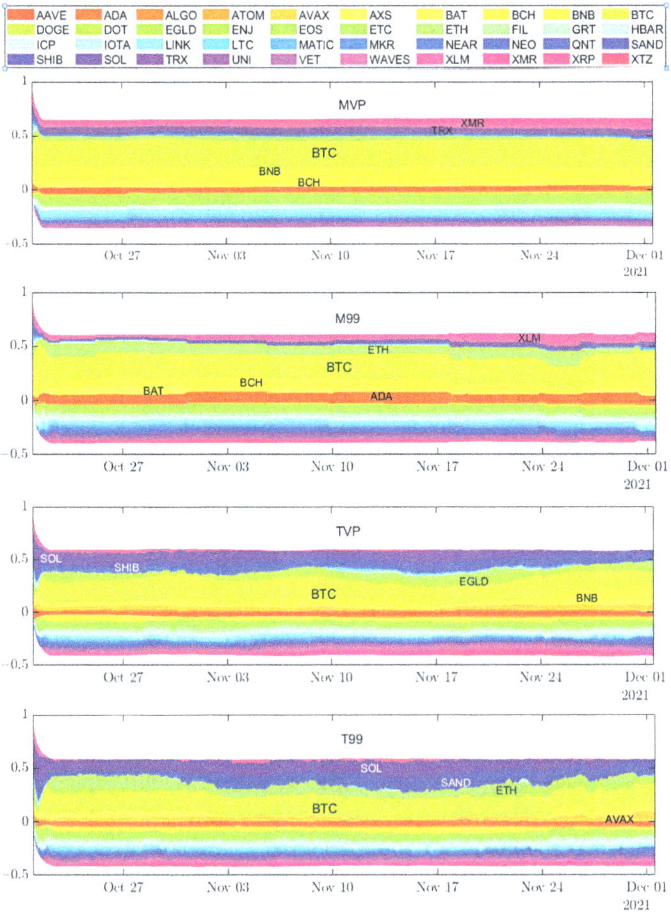

Fig. 4.8: Asset weights as a function of time for the MVP, TVP, M99 and T99 optimizations under the Lo–Patel strategy with $lev = 0.1$

In contrast to the positive weight holdings, where a few assets tend to dominate, under the constraints of the long–short strategies discussed here, no single asset can have a strong shorted position (large negative weight). This helps to minimize the risk of a large loss if the price of a shorted asset rises precipitously.

4.1.4 Momentum Strategy

The momentum strategy is based on the premise that investor sentiment tends to favor assets with positive performance (Tanous 1999; Soros 2015). This strategy is implemented by maintaining fixed portfolio weights over a predetermined holding period τ, during which no rebalancing occurs. Consequently, under a momentum strategy the portfolio

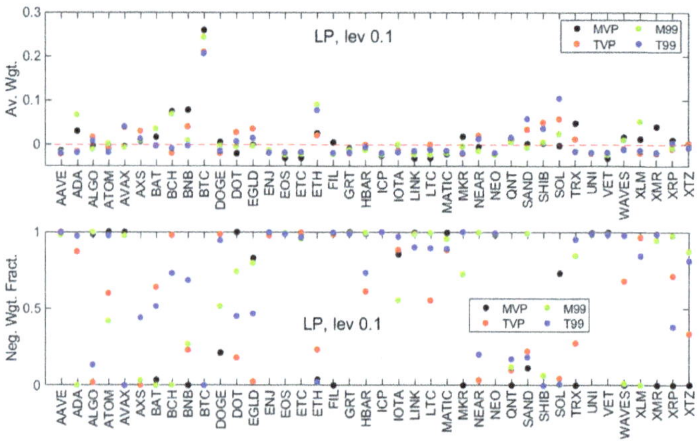

Fig. 4.9: (top) Average asset weight and (bottom) fraction of hourly time period an asset is shorted over the time period 22:00:00 on October 20, 2021 to 13:00:00 on December 1, 2021 for the indicated optimizations under the Lo–Patel strategy with $lev = 0.1$

return from time $t - \Delta t$ to t is no longer given by (2.6), but by

$$R_p(t) = \sum_{i=1}^{n} w_i(t_b(t))R_i(t),$$

where $t - \tau \le t_b(t) < t$ is the last time the weight values were rebalanced. Thus the portfolio return is a product of the latest asset returns with a vector of weights which may have been computed as many as τ time units ago. Thus while current returns may favor certain assets, these returns influence the portfolio return weighted by older weights which may, or may not, favor the same assets.

To display the strategy without extraneous considerations, we started with an initial portfolio that is optimized. Specifically we considered the Jacobs et al. strategy with $s = 0.1$. To determine the portfolio weights $w_i(t_0)$, we generated a transient sequence $w_i(k)$, $k = 1, \ldots, p$, with $w_i(1) = 1/n$, using under the $s = 0.1$ constraint (and any other imposed constraint) except that rebalancing was performed for each value of k. The weights $w_i(t_0)$ defined the initial optimized portfolio for the momentum strategy.

Fig. 4.10 shows the cumulative returns of the Jacobs et al. strategy constrained with $s = 0.1$ and rebalancing performed every $\tau \in \{1, 24, 120, 240\}$ hours (i.e. 1/24, 1, 5 and 10 days). The daily asset returns tend to determine where the peaks and valleys of the cumulative returns occur; the rebalancing period determines the strength of these extrema. Over the entire 1,000 hour (41 2/3 day) period, the T99 portfolio showed the best cumulative return performance under increased holding period, with a cumulative return on December 1, 2021 for $\tau = 240$ hours essentially equal to that for hourly rebalancing ($\tau = 1$). T95 revealed the second best ability to retain value over the full period as the τ increased. As τ increased, the TVP portfolio was unable to maintain the return levels achieved for $\tau = 1$. In contrast, the performance of the three minimum–risk portfolios

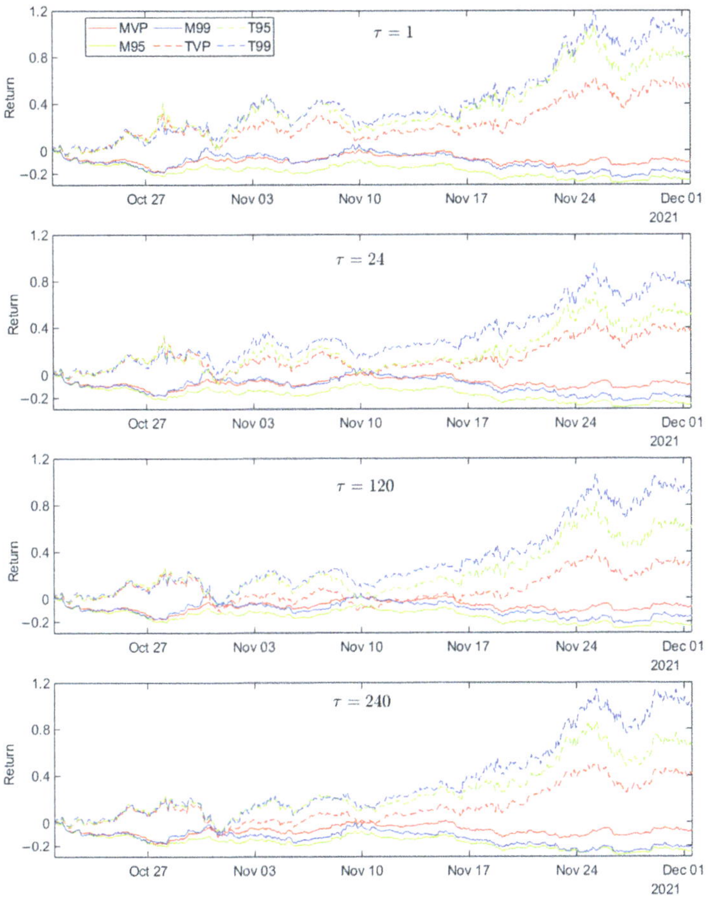

Fig. 4.10: Cumulative return for the Jacobs et al. $s = 0.1$ strategy with a holding period of $\tau = 1$, 24, 120 and 240 hours (1/24, 1, 5 and 10 days)

remained relatively constant with each increase in holding period. These results suggest that, with the exception of the TVP optimization, one can achieve a price-neutral effect by employing a momentum strategy with a reasonably long holding period, while benefiting greatly from reduced transaction costs.

Fig. 4.11 presents the weight profiles for the T99 portfolio for each of the holding periods. The effect of the holding period, even hourly, results in a staircase change in portfolio weights at rebalance times. Also shown are the time series of turnover values (characterizing a change in asset weights; see Section 4.2), which serve to highlight the (in)frequency of the rebalancing. In Section 4.2, we investigate how turnover constraints serve to smooth this staircase effect.

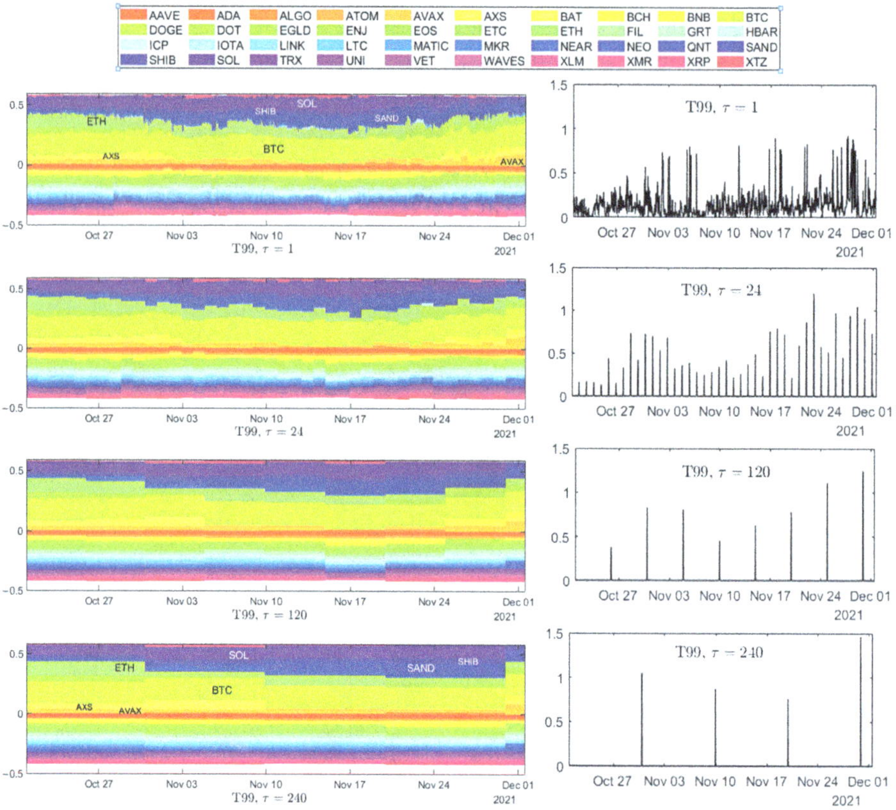

Fig. 4.11: (left) Asset weights and (right) turnover values as a function of time for the T99 optimization subject to the Jacobs et al. strategy with $s = 0.1$ and holding periods of $\tau = 1, 24, 120$ and 240 hours (1/24, 1, 5 and 10 days)

4.1.5 Minimum-Risk versus Tangent Portfolio Composition

Sections 4.1.1 – 4.1.4 display consistent composition differences between the minimum-risk and tangent portfolios. The minimum-risk portfolios were dominated by holding (as opposed to shorting) the crypto tokens BTC, BCH, BNB, XMR and TRX. BTC has the top market capitalization and is the most familiar crypto token; the blockchain BCH (see Chapter 1) was created as a hard fork of the BTC blockchain. BNB is the native token of the Binance exchange; XMR is a privacy focused crypto asset; and TRX is the native coin of the TRON network, one of the most widely used public blockchains.

The tangent portfolios were also long-dominated by BTC. In addition they concentrated holdings in SOL, SAND, SHIB, ETH, EGLD and AVAX. SOL runs a smart contract protocol; ETH is the second largest crypto token in terms of market capitalization – it introduced the functionality for smart contracts. EGLD is a platform capable of executing smart contracts extremely rapidly; AVAX also provides robust smart contract functionality. SAND (The Sandbox) is a Web3 decentralized game; while SHIB is a dog–themed (meme) token.

4.2 Performance Under Turnover Constraints

We address the issue of transaction costs arising from the rebalancing of asset weights. Since transaction costs are asset–specific, we employ turnover as a quantitative metric to assess the relative expense of transaction costs across different optimization strategies. Turnover is defined in terms of the L_1–norm of the change in asset weights from hour $t-1$ to t,

$$\|\Delta w_p(t)\|_1 = \sum_{i=1}^{n} |w_i(t) - w_i(t-1)|. \tag{4.4}$$

The turnover value is defined as half of the L_1–norm,

$$\mathrm{TO}(t) = \frac{1}{2}\|\Delta w_p(t)\|_1. \tag{4.5}$$

The total turnover of the portfolio over the time period $[0, T]$ is

$$\mathrm{TO}_{\mathrm{tot}}(T) = \sum_{k=1}^{T} \mathrm{TO}(k). \tag{4.6}$$

Assuming a common cost for both buy and sell transactions regardless of asset, the total turnover is proportional to the cost of maintaining the optimized weight structure of the portfolio over time.

Closely related[5] to the L_1–norm (4.4) is the L_2–norm,

$$\|\Delta w_p(t)\|_2 = \sqrt{\sum_{i=1}^{n} (w_i(t) - w_i(t-1))^2}\,. \tag{4.7}$$

Consideration of the L_2–norm in addition to $\mathrm{TO}(t)$ can provide discrimination among data points that share the same $\mathrm{TO}(t)$ value. (As we shall see in the case in which TO constraints are imposed.)[6]

Fig. 4.12 plots the time series of turnover values for the optimized portfolios under the long–only strategy in Section 4.1.1. In four of the plots, the turnover values are presented in log scale to show the wide range of hourly values. The M95 and M99 values

5 Consider the two vertices v_1 and v_2, connected by the hypotenuse of a right triangle. Intuitively, the L_2–norm (or L_2–distance), denoted as $\|v_1 - v_2\|_2$, represents the length of the hypotenuse, while the L_1–norm (or L_1–distance), denoted as $\|v_1 - v_2\|_1$, corresponds to the sum of the lengths of the other two sides of the triangle. Consequently, it follows that $\|v_1 - v_2\|_2 \leq \|v_1 - v_2\|_1$.

6 To continue the example of footnote 5, different hypotenuse lengths can be achieved for right triangles possessing side lengths that total to the the same sum.

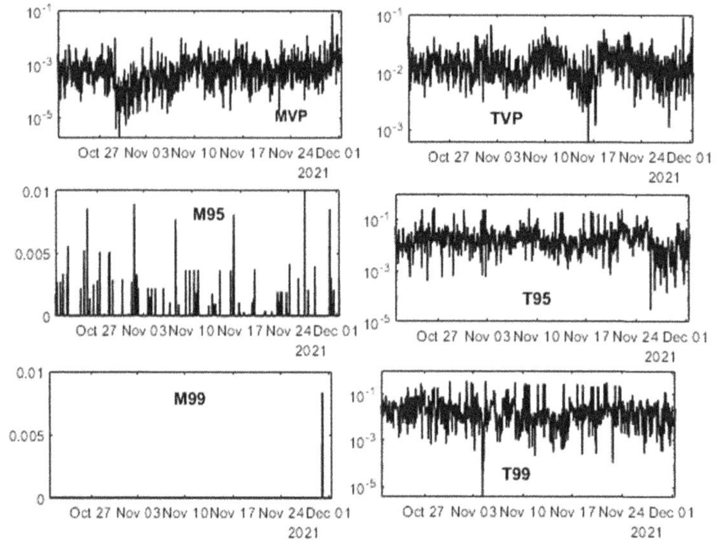

Fig. 4.12: Time series of turnover values (4.5) for the optimizations under the long–only strategy

are presented in linear scale. As shown in Fig. 4.2, the optimized M99 portfolio consisted of a single cryptocurrency (BTC) over this entire time period except for the 02:00:00 on November 30, 2021 optimization which added the crypto token SOL with a weight of 0.008. For M95, the optimized portfolio consisted of only two tokens, BTC and QNT, with infrequent weight changes between the two.

Box–whisker summaries provide a concise way to quantify the distributional information in a time series, as well as a convenient method to compare different time series. Fig. 4.13 presents box whisker summaries of the distributions of turnover values for the long–only, Jacobs et al. $s = 0.1$, and Lo–Patel $lev = 0.1$ strategies in Sections 4.1.1–4.1.3, respectively. The box–whisker summaries of the L_2 norm distributions are also presented. Note that the y–axis scale for the turnover and L_2 summaries are the same for the same strategy, but the y–axis scales differ between strategies. In all cases, the interquartile range (IQR) of the tangent portfolio is wider, and has larger (often much larger) Q_1, Q_2 and Q_3 values than the corresponding minimum–risk portfolio.

As noted previously, the constraints (4.3) for the Lo–Patel strategy are a form of constraint on the turnover, hence the cutoff of turnover distribution values at 0.1. In this case the L_2 norm summary statistics, which are not directly restricted, provide an enhanced comparison of outlier values for the Lo–Patel optimized portfolios. For the other two strategies, the range of outlier values above the upper whisker is better controlled under the long–only strategy than under the Jacobs et al, $s = 0.1$ strategy.

Fig. 4.13 also compares the total turnover (turnover values summed over the the period 22:00:00 on October 20, 2021 to 13:00:00 on December 1, 2021). Interestingly, for

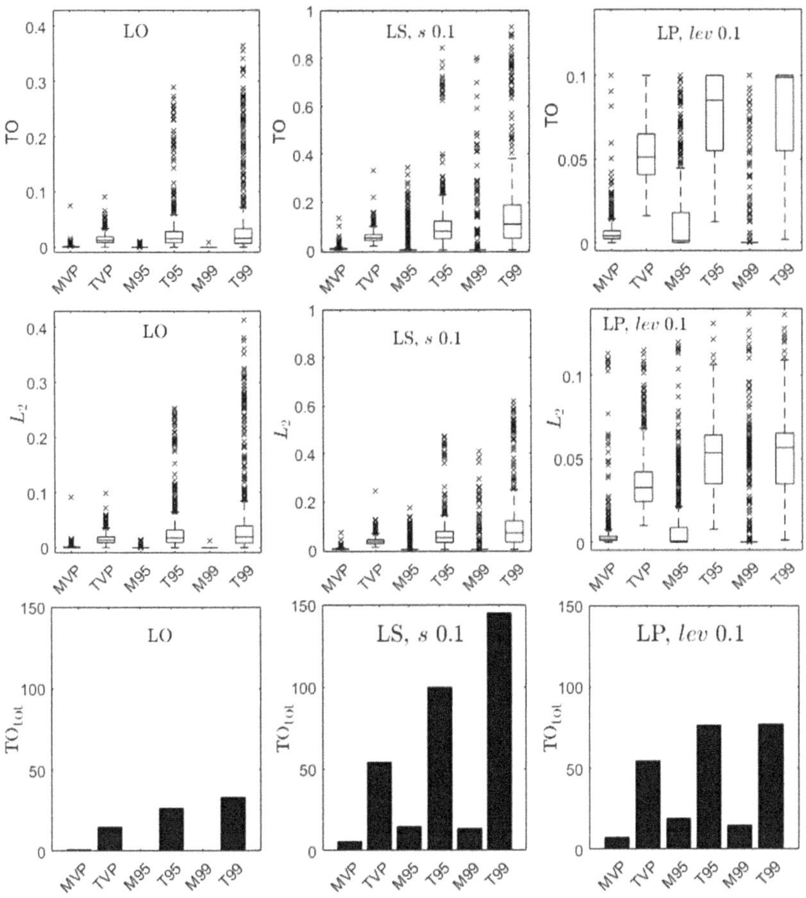

Fig. 4.13: Box-whisker summaries of (top) turnover, (middle) L_2 norm, and (bottom) total turnover values for the long-only (Section 4.1.1), Jacobs et al. $s = 0.1$ (Section 4.1.2) and Lo-Patel $lev = 0.1$ (Section 4.1.3) strategies

each of the six optimizations the long–only strategy was the cheapest. For each tangent portfolio, the Jacobs et al., $s = 0.1$, strategy was the most expensive.

The challenge lies in maintaining return performance while simultaneously reducing transaction costs. One strategy to achieve this is to implement a turnover constraint via the requirement

$$\mathrm{TO}(t) \leq C_{\mathrm{TO}} \,. \tag{4.8}$$

The value $C_{\mathrm{TO}} = \infty$ corresponds to no turnover constraint. Note that the turnover constraint can only be applied to the optimized portfolios. By their definition, the two benchmark portfolios, EWP and EWBH, can not be subjected to further constraints. In addition, the Lo–Patel strategy constraints (4.3) provide a natural turnover restriction. We therefore consider adding the constraint (4.8) to the long–only and Jacobs et al. strategies.

Instead of enforcing a strict turnover constraint as in (4.8), recent studies (Broadie and Glasserman 1996; DeMiguel et al. 2009; Yen 2016) have introduced L_1– or L_2–norm constraints on portfolio–weight changes by incorporating penalty terms into the optimization function. This approach enhances portfolio stability by promoting sparsity (i.e., increasing the number of zero changes in portfolio weights, thereby reducing the number of active weights) and mitigating overfitting.[7]

Fig. 4.14 presents the cumulative return performance for the six optimized portfolios under the long–only strategy with $C_{TO} = \infty$, 0.04, 0.00134 and 0.000114. The choice $C_{TO} = 0.04$ allows for the possibility of 100% turnover in the asset weights in 25 hours. The choices $C_{TO} = 0.00134$ and 0.000114 correspond, respectively, to the possibility of 100% asset weight turnover in 31 and 365 days. The 4% turnover constraint is very liberal, while the 0.0114% constraint is very stringent. Also shown is the cumulative return of the EWBH benchmark portfolio. (Recall the the EWBH portfolio incurs no turnover and hence can be maintained with no transaction costs.)

The return performance of the tangent portfolios remains relatively consistent for $C_{TO} \in \{\infty, 0.04, 0.00134\}$. While their performance degraded under the 0.0114% constraint, they still outperformed the EWBH benchmark. In contrast, the minimum–risk long–only portfolios show virtually unchanged cumulative return behavior between themselves and over the range of turnover constraints. All minimum–risk portfolios underperformed the EWBH benchmark, and lost just under 14% of their value over the 1,000 hours of simulation.

Fig. 4.15 shows the asset weight composition for the T99 portfolio as the turnover constraint C_{TO} value is decreased. Two trends are apparent. The first, to be expected, is that the rate of change of asset weight "became smoother" as C_{TO} decreased (i.e. as the turnover constraint becomes more stringent). The second is that the T99 portfolio composition became increasingly concentrated in two assets (SOL and SHIB); the diversity of the portfolio decreased. In contrast, as documented for MVP in Fig. 4.16, the minimum–risk portfolios were overwhelmingly concentrated in BTC, so that the cumulative return performance of all minimum–risk portfolios is essentially that of BTC. As a consequence, all three minimum–risk cumulative return time series are essentially identical, and independent of the value of C_{TO} value (which restricts weight changes and not weight values).

Fig. 4.17 displays the box–whisker summaries of turnover values of the optimized portfolios under the long–only strategy with $C_{TO} = \infty$, 0.04, 0.00134 and 0.000114. Note the decrease in the range of the y–axis scale with the decrease in C_{TO}. As the M99 portfolio consists exclusively of the BTC asset its turnover was zero (with the exception of two hours, 02:00:00 and 03:00:00 on November 30, 2021 when a small weighting of SOL was added, causing two outlier values in turnover). The liberal 4% constraint eliminates

7 Overfitting occurs when the optimization process excessively tailors the portfolio to fit the random noise component in the financial data.

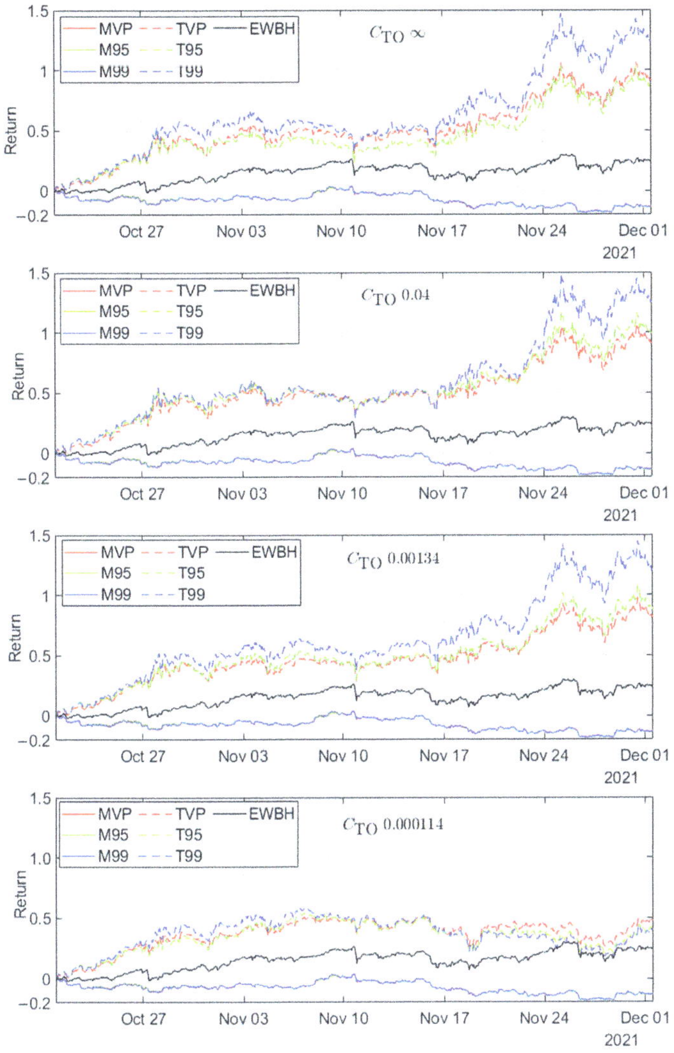

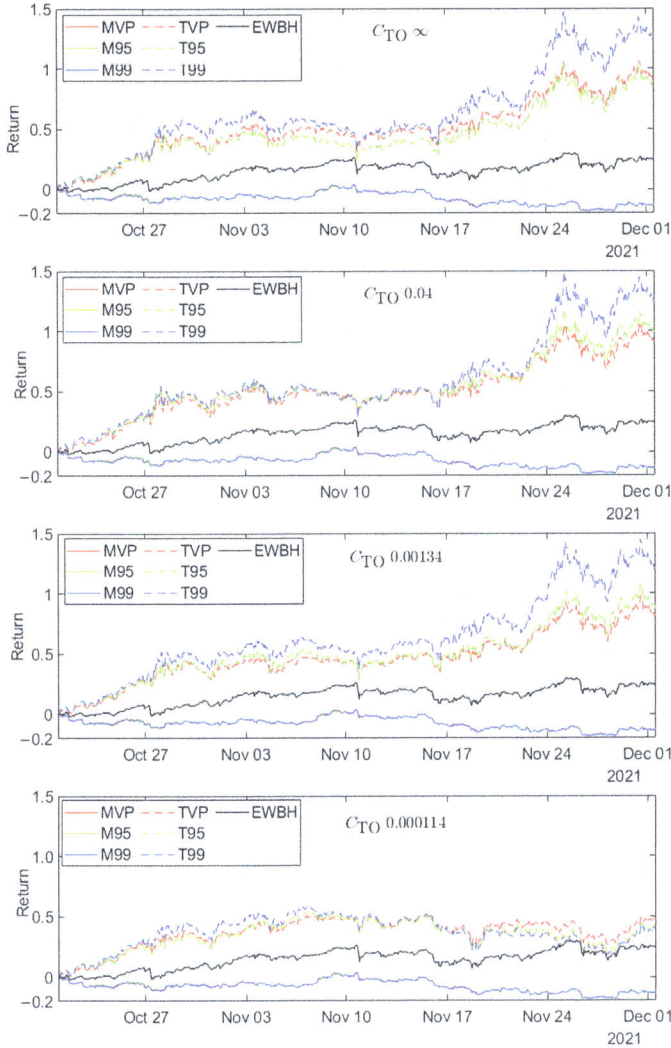

Fig. 4.14: Cumulative return for the optimizations under a long-only strategy with $C_{TO} = \infty$, 0.04, 0.00134 and 0.000114. The cumulative return for the EWBH portfolio is included

outlier values for the T95 and T99 portfolios. It reduces the number of outliers for the TVP and MVP portfolios. The stringent 0.0114% constraint limits every hourly turnover for the three tangent portfolios, and virtually every hourly turnover for M95; for MVP, there are a number of hourly turnovers that fall below the 0.0114% limit. The effects of the 0.134% turnover constraint can be similarly "read" from the corresponding box–whisker plot.

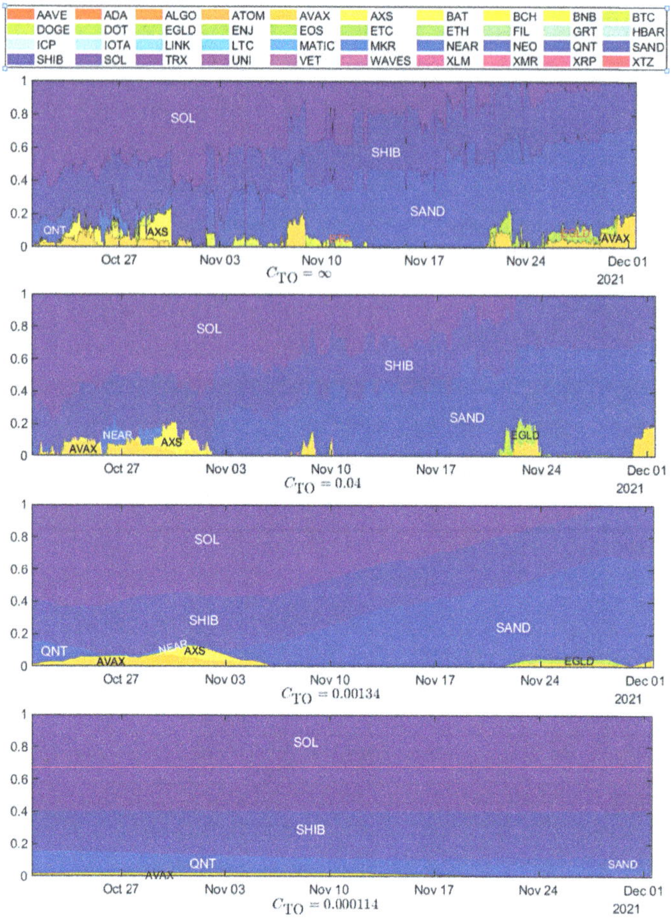

Fig. 4.15: Asset weights as a function of time for the T99 optimization under a long-only strategy with $C_{TO} = \infty$, 0.04, 0.00134 and 0.000114

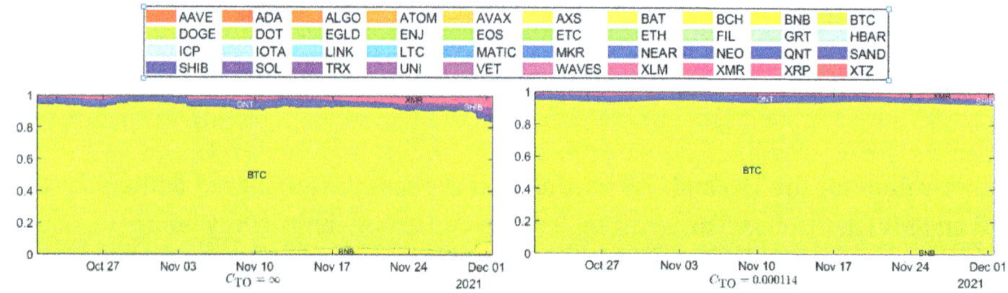

Fig. 4.16: Asset weights as a function of time for the MVP optimization under a long-only strategy with $C_{TO} = \infty$ and 0.000114

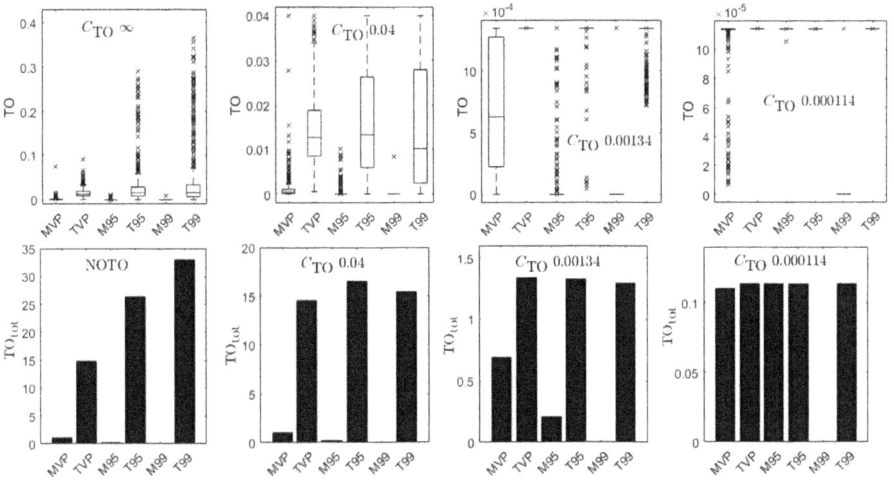

Fig. 4.17: Box–whisker summaries of turnover (TO) and total turnover (TO_{tot}) values for the optimizations under the long–only strategy with $C_{TO} = \infty$, 0.04, 0.00134 and 0.000114

Fig. 4.17 also summarizes the total turnover cost for the 1,000 hour simulation period for each optimized portfolio under the different turnover constraint values. Compared to no turnover constraint, the total cost of the T95 and T99 portfolios were somewhat better controlled under the 4% constraint. Again (including the trivial case of the M99 portfolio for the reason noted above), the transaction cost of each portfolio was markedly reduced under the 0.0114% constraint. The intermediate 0.134% constraint limited the total transaction cost of the tangent portfolios, but not the minimum–risk ones. Assessing the results of Figs. 4.15 – 4.17, one might consider opting for the T99 portfolio under a long–only, $C_{TO} = 0.00134$, constraint as a compromise between return maximization and transaction cost reduction.

As noted in Section 4.1.4, extending the holding period can produce a temporal staircase change in asset weights (See Fig. 4.11). Imposition of a turnover constraint smooths the staircasing. Fig. 4.18 presents the weight profiles for the T99 portfolio for each holding period, $\tau = 1$, 24, 120 and 240 hours under the Jacobs et al. strategy with $s = 0.1$ and $C_{TO} = 0.04$. Also shown are the time series of turnover values. For $\tau = 1$ hour, the total weight change is restricted by the maximum limit, $C_{TO} = 0.04$, approximately half of the time. For $\tau = 24$, 120 and 240 hours, the turnover constraint "pins" every total weight change at the maximum limit. The resulting weight profiles are more smoothly changing than for those displayed in Fig. 4.11.

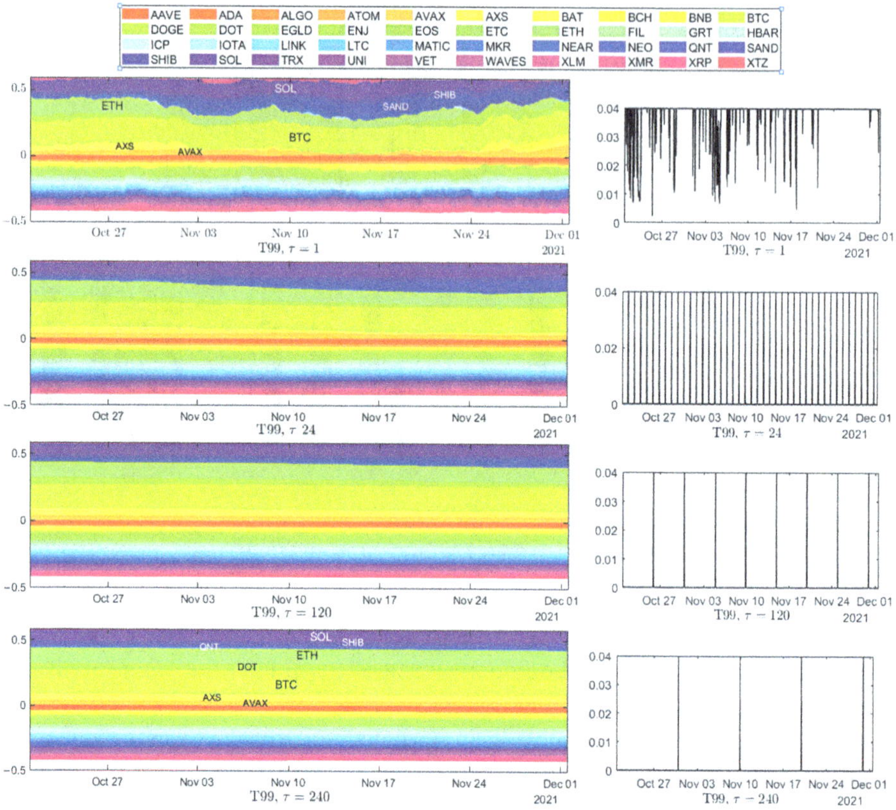

Fig. 4.18: (left) Asset weights and (right) turnover values as a function of time for the T99 optimization subject to the Jacobs et al. strategy with $s = 0.1$, $C_{TO} = 0.04$, and holding periods of $\tau = 1, 24, 120$ and 240 hours (1/24, 1, 5 and 10 days)

4.3 Performance–Risk Measures

In Sections 4.1 and 4.2, we evaluate portfolio performance using cumulative return metrics and turnover. While these metrics are crucial, they need to be weighed against a wide array of alternative performance measures. Cogneau and Hübner (2009) propose a classification framework encompassing 101 performance measures, organized into four categories: performance–risk ratios, incremental return, investor risk preferences (expressed through a utility function), and market timing. The most extensive category is performance–risk ratios, which Cogneau and Hübner (2009) further divide into three subcategories based on whether the risk component is absolute, systematic, or nonsystematic (i.e., diversifiable). Factor models, such as Jensen's alpha (Jensen 1968) and the Fama–French three– and five–factor models (Fama and French 1993, 2015), fall under the incremental return category. From the performance–risk class, we consider four measures.

1. Maximum drawdown (MDD):

$$\text{MDD}(T) = \max_{t\in(t_1,t_T)} \left[\max_{s\in(t_1,t)} \left(\frac{P_p(s) - P_p(t)}{P_p(s)} \right) \right], \tag{4.9}$$

where $P_p(s)$ is the price of the portfolio at time s. MDD(T) denotes the largest peak-to-trough decline in the portfolio price over the time period (t_1, t_T). Expressing MDD as a fraction (equivalently percentage) is critical, as a 1M BUSD drop from a peak is (relatively) insignificant if the peak value is 100M BUSD, whereas it is incredibly important if the peak value is 1.1M BUSD. Note that (4.9) can equivalently be computed using cumulative return $(R_p(t_1, s) - R_p(t_1, t)/R_p(t_1, s))$, rather than price.

2. The Sharpe ratio (Sharpe 1966):

$$\text{SR}(T) = \frac{\text{E}[r_p(t) - r_f(t)]_{[t_1,t_T]}}{\sqrt{\text{Var}[r_p(t) - r_f(t)]_{[t_1,t_T]}}} = \frac{\mu_{p[t_1,t_T]}}{\sigma_{p[t_1,t_T]}}, \tag{4.10}$$

where $r_p(t)$ is the portfolio return, $r_f(t)$ is the risk-free rate at time t, and μ_p and σ_p are, respectively, the expected mean and standard deviation of the portfolio's excess return, $r_p(t) - r_f(t)$. The numerator of the Sharpe ratio does not distinguish between gains, $r_p(t) > r_f(t)$, or losses, $r_p(t) < r_f(t)$; it includes information from both tails of the return distribution. The same statement is true of its denominator. By measuring overall mean (numerator) and volatility (denominator) the Sharpe ratio measures two parameters characteristic of the "center" of a return distribution.

3. The Sortino–Satchell ratio (Sortino and Satchell 2001):

$$\text{SS}_p(T) = \frac{\text{E}\left[(r_p(t) - r_f(t))^+\right]_{[t_1,t_T]}}{\| (r_f(t) - r_p(t))^+ \|_{p[t_1,t_T]}}, \tag{4.11}$$

where $y^+ = \max(0, y)$. The Sortino–Satchell ratio is defined for a general p-norm in the denominator; we specifically choose $p = 2$. For brevity we refer to the $p = 2$ choice as the Sortino–Satchell ratio.[8] The Sortino–Satchell ratio addresses the concern mentioned above regarding the Sharpe ratio. The numerator of the Sortino–Satchell ratio measures average positive gain while the denominator measures the spread of loss. In this sense it is more of a true reward:risk measure than is the Sharpe ratio.

4. The Rachev ratio (Rachev et al. 2008):

$$\text{RR}_{\alpha,\beta}(T) = \frac{\text{CVaR}_\alpha(r_f(t) - r_p(t))_{[t_1,t_T]}}{\text{CVaR}_\beta(r_p(t) - r_f(t))_{[t_1,t_T]}}, \tag{4.12}$$

which represents the reward potential for positive returns compared to the risk potential for negative returns at quantile levels α and β defined by the user. Thus

8 When $p = 2$, the Sortino–Satchell ratio is also known as the Sortino ratio.

it provides a measure that concentrates on the "far" tail regions to a greater extent than does the SS ratio. In our analysis, we set $\alpha = \beta = 0.95$ and will refer to it simply as the Rachev ratio. This ratio is well-defined provided that the numerator and denominator are strictly positive (Cheridito and Kromer 2013).

Note that MDD is a risk measure based upon price, while the three ratio measures are based upon returns. A smaller value of MDD indicates better performance; for the ratios, higher values signify better performance. The MDD has the following property. Consider a peak value occurring at time t_1 and trough occurring at time $t_2 > t_1$. Consider a moving window of size $T > t_2 - t_1$. Then this peak–to–trough price difference will be present in all moving windows $[t - T, t]$ for $t = t_2$ to $t = t_1 + T$. Assume this peak–to–trough price difference is a maximum drawdown for time period T. Then MDD(T) will remain unchanged in value over the period $t \in [t_2, t_1 + T]$. The larger the value of T, the longer MDD(T) will remain unchanged.

The selection of these performance–risk ratios was guided by their representation of various subclasses within the Cogneau–Hübner classification and by the work of Cheridito and Kromer (2013), who examined the properties of reward–risk measures in relation to four desirable qualities: monotonicity, quasi–concavity, scale invariance, and whether the measure is distribution-based. They argue that every performance measure should at least be monotonic (where "more" is preferable to "less") and quasi–concave (favoring averages over extremes and encouraging risk diversification rather than concentration). The Sortino–Satchell ratio (SS) satisfies all four properties. The Sharpe ratio (SR), the most widely used performance measure, meets three of the four properties; it does not guarantee monotonicity, which is arguably the most critical property of a risk measure. The Rachev ratio (RR), often used by hedge funds aiming for high returns while insuring against significant losses, also satisfies three of the four properties; it does not ensure quasi–concavity.

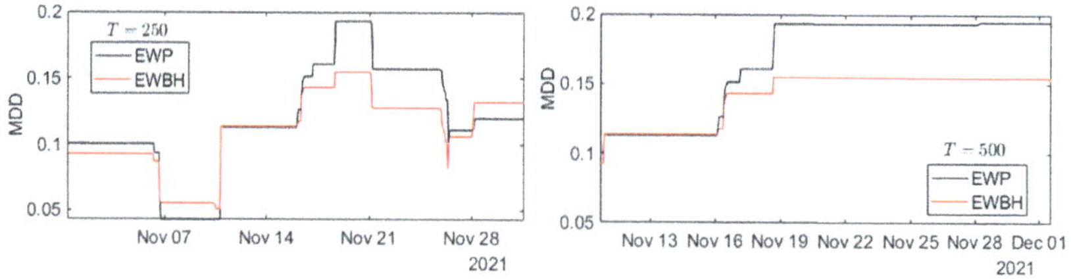

Fig. 4.19: MDD for the EWP and EWBH using $T = 250$–hour and $T = 500$–hour moving windows

In our analysis, we calculated values for each performance measure using a moving window of length T, providing a time–series perspective on the evolving measure. To select T, we analyzed the MDD of the EWP and EWBH portfolios. Figure 4.19 illustrates the MDD plots for both portfolios using $T = 250$ and $T = 500$ hours. The time periods covered are from 6:00:00 on October 31, 2021 to 13:00:00 on December 1, 2021 for the 250–hour window, and from 16:00:00 on November 11, 2021 to 13:00:00 on December 1, 2021 for the 500–hour window. For a fixed window length, the MDD values for the EWP and EWBH display similar trends. The results highlight the memory effect of MDD under longer values of T discussed above. The 500–hour moving window smooths too much detail, obscuring important short–term variations. Consequently, to retain more granular insight, we proceed with further comparisons of the performance measures using the 250–hour moving window.

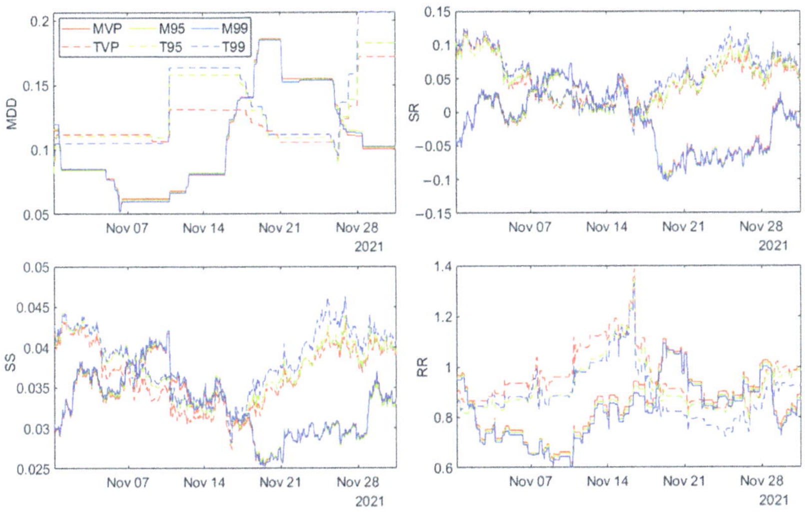

Fig. 4.20: MDD, SR, SS, and RR time series for the optimizations under the long–only strategy with C_{TO} = 0.00134

Fig. 4.20 displays the performance–measure time series for the long–only cryptocurrency portfolios with C_{TO} = 0.00134, having the cumulative returns presented in Fig. 4.14. Note that the SR and SS trends were similar to each other, while the MDD and RR trends showed similar behavior (but with opposing conclusions in that, when the MDD for minimum–risk portfolios was best, RR for minimum–risk portfolios was worst). For the majority of the time period, the tangent portfolio MDD values were worse than those for the minimum–risk portfolios, while the SR, SS and RR values of the tangent portfolios were better than the minimum–risk ones. In general, tangent portfolios are more aggressive at enhancing reward:risk measures, but at the cost of greater drawdowns under market reversals.

Reflecting their predominant weighting of BTC, the MDD values for the three minimum–risk portfolios were essentially identical; essentially measuring the MDD values for BTC. The same is true for the SR and SS values. The tail measure RR reveals the greatest difference between the minimum–risk portfolios; RR is more sensitive to the presence of small admixtures of other assets. As the M99 portfolio consists 100% of BTC (with the exception of two hours as noted) while MVP has the greatest (though still small) admixture of other assets (mostly QNT, TRX, SHIB and XMR), close examination of the RR time series shows MVP had larger values than M95 which, in turn, had large values than M99. The small admixture of other assets in addition to BTC increased the RR value.

The risk measure values for the three tangent portfolios exhibited noticeable differences between them but all followed similar trends. Of the three, the TVP portfolio generally had the smallest (best) MDD values; T99 had the largest (worst). This relationship held true for RR, with TVP having the largest (best) values and T99 the smallest (worst). For SR and SS, this was reversed, with T99 having the largest (best) values, and TVP the smallest (worst).

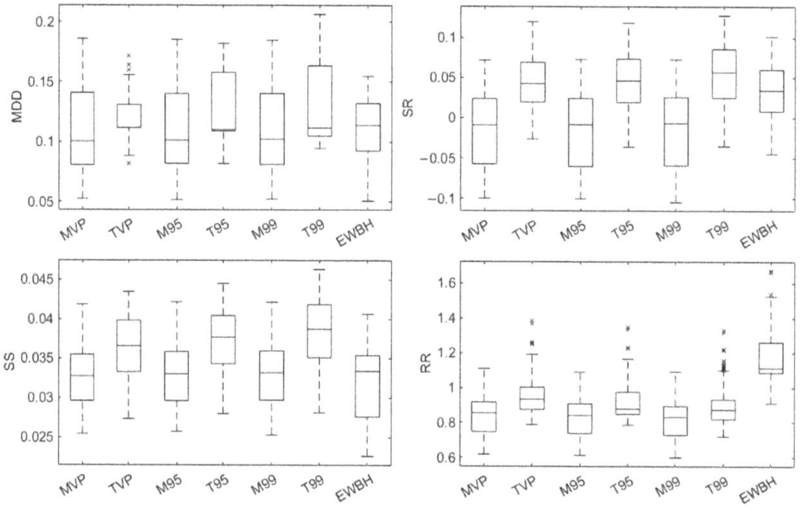

Fig. 4.21: Box–whisker summaries of the statistics of the performance measures for the long–only strategy with $C_{TO} = 0.00134$. The summaries for the benchmark EWBH are also included

While box–whisker statistical summaries of a distribution of time series values result in a loss of temporal granularity, they provide a clearer perspective on the quantiles and outliers for each method. Fig. 4.21 contrasts the MDD, SR, SS, and RR summary statistics for the six optimized portfolios under the long–only strategy with $C_{TO} = 0.00134$. Also shown are the risk measure summaries for the benchmark EWBH. For the reasons already discussed, the box whisker summaries for the minimum–risk portfolios are virtually identical to that of BTC. For MDD, the box–whisker plots indicate that the minimum–risk portfolios were preferable to their tangent portfolio counterparts, although TVP is, in some aspects, competitive with MVP. The conclusions are opposite under SR, SS, and RR where the summary statistics were better for the tangent portfolios, and worse for their minimum–risk counterpart.

The benchmark EWBH presented an interesting risk measure character. Under MDD, its statistical distribution was only slightly worse than for the minimum–risk portfolios, in that it had higher Q_1 and Q_2 values. However its Q_3 and upper whisker values where better than any of the minimum–risk portfolios. Under SR it was only slightly worse than the tangent portfolios, but significantly better than the minimum–risk portfolios. Interestingly, under SS its distribution is comparable to, and even slightly worse than, the minimum–risk portfolios. Under RR, its distribution is superior to all optimized portfolios. These empirical observations form part of the evidence that diversified, buy–and–hold portfolios often perform very well.

Using total risk measure values, i.e. the calculation of a single value for each performance metric across the entire 751–hour period from 06:00:00 on October 31, 2021 to 13:00:00 on December 1, 2021,[9] enables rapid comparison among many portfolio strategies. Fig. 4.22 compares total risk measure values of the optimized portfolios under the long–only strategy with $C_{\mathrm{TO}} = \infty$, 0.04, 0.00134 and 0.000114. Use of the total risk measure shows distinct differences between SR and SS measures. It also “picks up” the subtle differences between the minimum–risk portfolios by capturing the total effect of the differences in small admixtures of assets other than BTC in these portfolios.

The benchmark EWBH had the best overall MDD and RR values. Interestingly, TVP had a better overall MDD value than MVP, while T95 bested M95 in MDD under three of the turnover constraint values. Under overall SR and SS, the tangent portfolios always outperformed the minimum–risk portfolios. The overall Sharpe ratio of the minimum–risk portfolios was always negative; for the tangent portfolios and EWBH it was always positive. Under RR, TVP and T95 always outperformed their minimum–risk counterpart.

Fig. 4.23 provides the total risk measure summary for the optimized portfolios under the Jacobs et al. strategy with $s = 0.1$ and $C_{\mathrm{TO}} = 0.04$ under change of holding period, $\tau = 1$, 24, 120 and 240 hours. The risk measures for the tangent portfolios were more sensitive (got worse) to an increase in holding period compared to the minimum–risk portfolios.

It is an undeniable conclusion from the results in this chapter that different performance measures produce different rankings of portfolio performance. While price (cumulative return) performance is always a primary concern, other performance measures become important when the market becomes volatile or enters a downturn. Thus monitoring a portfolio under a variety of performance measures remains a crucial management component.

9 This matches the time period covered by the time series computations of Figs. 4.20 and 4.21.

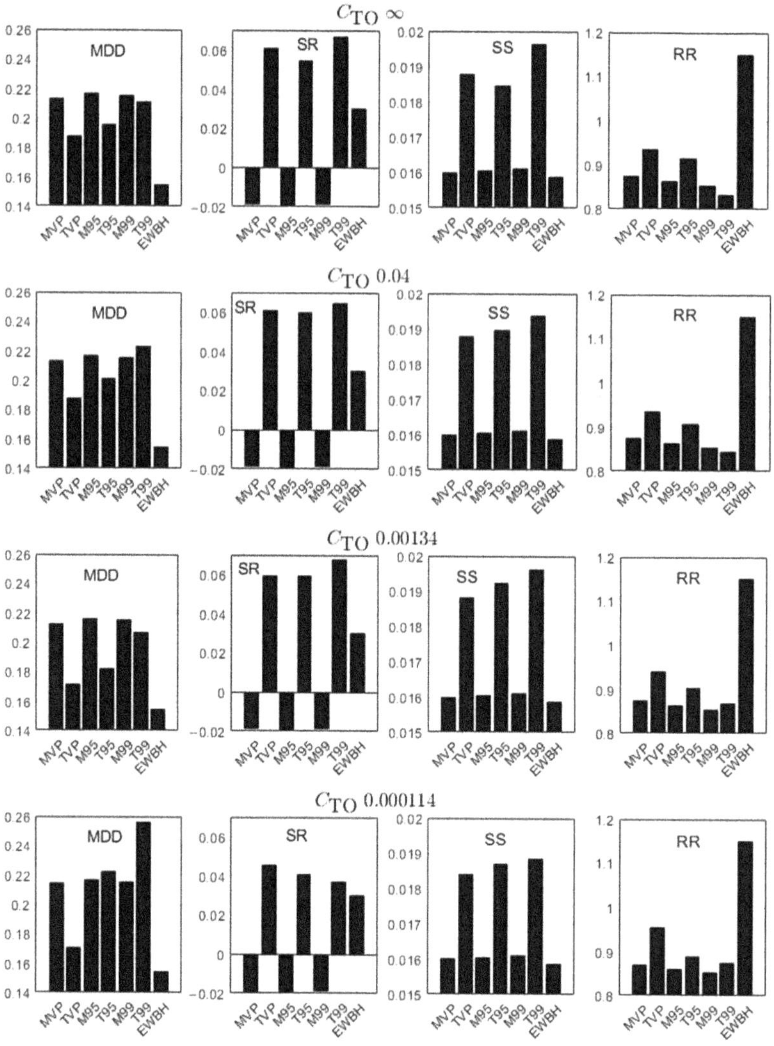

Fig. 4.22: Total risk measure values of the optimizations under the long-only strategy with $C_{TO} = \infty$, 0.04, 0.00134 and 0.000114

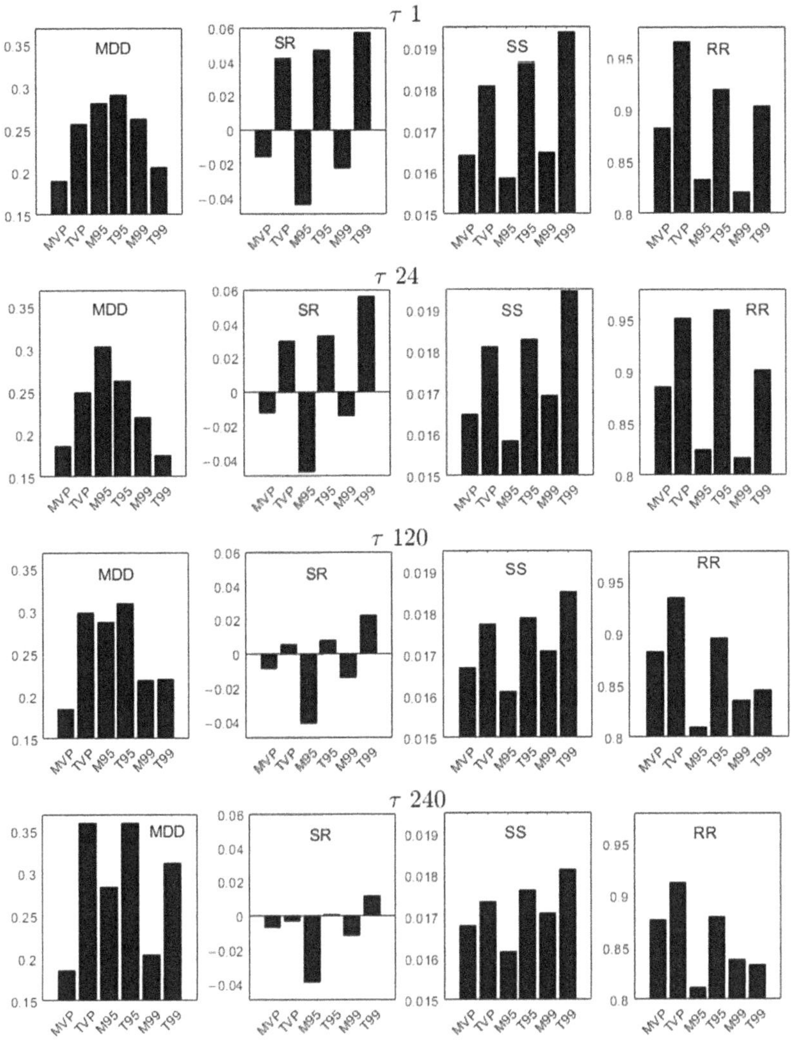

Fig. 4.23: Total risk measure values of the optimizations under the Jacobs et al. strategy with $s = 0.1$, $C_{TO} = 0.04$, and holding periods of $\tau = 1, 24, 120$ and 240 hours.

5 Dynamic Portfolio Optimization

The optimization approach detailed in Chapter 4 involves sequentially sampling return data through moving windows over a predetermined historical period that encompasses a limited set of market activities and global events. As is commonly stated in investment fund prospectuses, "past performance is not necessarily indicative of future results". In contrast to merely relying on historical asset return samples, dynamic optimization assumes that historical returns consist of samples from a dynamic multivariate return distribution. Each asset return is a sample from the corresponding marginal distribution of the multivariate distribution. Dynamic optimization aims to uncover this multivariate distribution,[1] which, once known, can be used to produce extensive predictive samples of correlated asset returns that better capture the distribution's tail behavior (i.e., extreme market events). This results in a portfolio optimization that is more sensitive to potentially significant market shifts.

The dynamic optimization method we discuss consists of the following components. Within each moving window, apply an ARMA(1,1)–GARCH(1,1) model to the return data of each asset. This choice, from a general ARMA(p,q)–GARCH(m,n) model, provides sufficient flexibility while limiting to a reasonable number of parameters. To model the innovations in the ARMA–GARCH framework, use a Student's t–distribution, which accommodates heavier tails than the (commonly used) normal distribution. The innovations from these marginal distribution fits are then transformed into copula space.[2] The (empirical) copula transformation is critical, as in copula space all parts of the distribution (specifically, for our concern, the tail regions) are equally weighted. The copula–transformed innovations are then fit to a multivariate t–copula distribution, capturing the interdependencies among asset innovations. A large synthetic sample of asset innovation values is generated from this t–copula distribution. After reversing the copula transformation on this sample, these innovations are used in the ARMA–GARCH model to produce a corresponding large synthetic sample of portfolio asset–return scenarios. These scenarios are then input into the optimization algorithm to determine the portfolio's asset weights for the subsequent hour.

We delve into the specifics of each step in this optimization process in Sections 5.1. In Section 5.2, we demonstrate dynamic optimization of the cryptocurrency portfolio under long–only and long–short strategies.

1 In practice, a parametric approximation to the distribution.

2 The space determined by the multivariate cumulative distribution function.

https://doi.org/10.1515/9781501517136-005

5.1 Development of the Dynamic Optimization Method

Consider n assets, with asset $i = 1, \ldots, n$ having mean return $\overline{r}_i$. Consider a moving window that spans T time units, encompassing nT asset return values $r_{i,t}$, $i = 1, \ldots, n$, $t = 1, \ldots, T$. While $\widehat{\mu}_i = T^{-1}\sum_{t=1}^{T} r_{i,t}$, serves as an estimate for $\overline{r}_i$ with an error of $O\left(T^{-1/2}\right)$, the estimates for the covariance matrix components $\Sigma_{ij} = \sum_{t=1}^{T}(r_{i,t} - \widehat{\mu}_i)(r_{j,t} - \widehat{\mu}_j)$ tend to have larger error. This is because the same set of values $r_{i,t} - \widehat{\mu}_i$, $t = 1, \ldots, T$, is reused to calculate the values for the set of components Σ_{ij}, $j = 1, \ldots, n$, $j \neq i$. Thus, for any realistically sized moving window, there is a risk of inadequate sampling in the computation of the covariances observed between assets. This observation is clearly applicable to mean–variance optimization. For CVaR–based optimization, it is important that the historical return sample provide adequate sampling of the tail regions of the multivariate distribution. In practice, this is never the case. Dynamic optimization aims to provide a statistically robust large sample with adequate sampling in the tail regions.

5.1.1 ARMA(1,1)–GARCH(1,1) with Student's *t*-Distribution

If an asset's time series of returns r_t[3] is stationary, a commonly used parametric model to describe it consists of a combination of the autoregressive moving–average (ARMA) and the generalized autoregressive conditional heteroskedasticity (GARCH) models. The ARMA component (Engle 1982) explicitly models the behavior of the mean trend of the returns, while the GARCH component (Bollerslev 1986) models the variance. The ARMA(p, q) model is (Tsay 2010)

$$r_t = \phi_0 + \sum_{i=1}^{p} \phi_i r_{t-i} + a_t + \sum_{j=1}^{q} \theta_j a_{t-j}, \tag{5.1}$$

where each shock a_t is a zero–mean random variable. The second term in (5.1) represents the autoregressive dependence of r_t on past returns; the last two terms incorporate the effect of a weighted moving average of the shocks. The GARCH(m, s) model relates a_t to the variance σ_t^2 of the series (Tsay 2010),

$$\begin{aligned} a_t &= \sigma_t \epsilon_t, \\ \sigma_t^2 &= \alpha_0 + \sum_{i=1}^{m} \alpha_i a_{t-i}^2 + \sum_{j=1}^{q} \beta_j \sigma_{t-j}^2. \end{aligned} \tag{5.2}$$

Here, the innovations ϵ_t are zero–mean, unit–variance, iid random variables. The GARCH model is autoregressive in both σ_t^2 and a_t^2.

3 We simplify the asset return notation from $r_i(t)$ to r_t in this section to present ARMA–GARCH using standard time series notation.

The ARMA(1,1)–GARCH(1,1) model has six parameters,

$$r_t = \phi_0 + \phi_1 r_{t-1} + a_t + \theta_1 a_{t-1}, \tag{5.3a}$$

$$a_t = \sigma_t \epsilon_t, \tag{5.3b}$$

$$\sigma_t^2 = \alpha_0 + \alpha_1 a_{t-1}^2 + \beta_1 \sigma_{t-1}^2, \tag{5.3c}$$

providing sufficient flexibility to model the return time series of different assets. Fitting a specific time series requires specifying the distribution that governs the innovations ϵ_t of the random variables. We assume that the innovations follow the Student's t–distribution

$$t_\nu(x) = \frac{\Gamma\left((1+\nu)/2\right)}{\sqrt{\nu\pi}\,\Gamma(\nu/2)}\left(1 + \frac{x^2}{\nu}\right)^{-(1+\nu)/2}, \tag{5.4}$$

where $\Gamma(\cdot)$ is the gamma function.[4] This distribution is symmetric and has fatter tails compared to the normal distribution. The parameter ν is known as the degrees of freedom. The distribution has finite variance for $\nu > 2$, zero skewness for $\nu > 3$, and finite kurtosis for $\nu > 4$. For values of ν less than these thresholds, these moments are either infinite or undefined. As ν approaches infinity, the Student's t–distribution converges to the normal distribution. (In practice, the normal distribution is a good approximation when $\nu \geq 30$).

The cumulative distribution function (CDF) for the Student's t–distribution is

$$T_\nu(x) = \frac{1}{2} + x\Gamma\left(\frac{\nu+1}{2}\right)\frac{{}_2F_1\left(\frac{1}{2}, \frac{\nu+1}{2}; \frac{3}{2}; -\frac{x^2}{\nu}\right)}{\sqrt{\pi\nu}\,\Gamma\left(\frac{\nu}{2}\right)} \tag{5.5}$$

where ${}_2F_1(a, b; c; z)$ is the ordinary hypergeometric function.

5.1.2 Multivariate *t*–Distribution and *t*–Copulas

The Student's t–distribution (5.4) is applicable to univariate random variables. For random vectors, whose elements may be correlated, the appropriate extension is a multivariate t–distribution. Let $\boldsymbol{X}$ denote an n–variate, t–distributed random variable having realized values (n–vectors) denoted by $\mathbf{x}$. There are various forms of a multivariate t–distribution; we employ the commonly used version,

$$t_\nu(\boldsymbol{x}; \boldsymbol{\mu}, \Sigma) = \frac{\Gamma\left(\frac{(n+\nu)}{2}\right)}{(\nu\pi)^{n/2}\Gamma(\frac{\nu}{2})\sqrt{|\Sigma|}}\left(1 + \frac{(\boldsymbol{x}-\boldsymbol{\mu})^\top\Sigma^{-1}(\boldsymbol{x}-\boldsymbol{\mu})}{\nu}\right)^{-(n+\nu)/2}. \tag{5.6}$$

We assume $\boldsymbol{X} \sim t_\nu(\boldsymbol{x}; \boldsymbol{\mu}, \Sigma)$.[5] The parameters of this model are Σ, $\boldsymbol{\mu}$, and ν. A notable feature of this version of the multivariate t–distribution is that ν remains a scalar mea-

4 The gamma function $\Gamma(x)$ is defined as $\Gamma(x) = \int_0^\infty t^{x-1}e^{-t}\,dt$, which satisfies the recursion relation $\Gamma(x+1) = x\Gamma(x)$.

5 The notation $X \sim t_\nu(\cdot)$ indicates that the random variable X is distributed according to the probability distribution $t_\nu(\cdot)$.

sure of the degrees of freedom.[6] In (5.6), $\boldsymbol{x}$ has mean value $\boldsymbol{\mu}$ (for $\nu > 1$).[7] The covariance matrix of $\boldsymbol{x}$ is proportional to the $n \times n$ matrix Σ; specifically the covariance is $\nu\Sigma/(\nu-2)$ for $\nu > 2$. The notation $|\Sigma|$ represents the determinant of Σ.

The CDF for this multivariate distribution is defined as

$$T_\nu(\boldsymbol{y}; \boldsymbol{\mu}, \Sigma) = \Pr(\boldsymbol{X} \leq \boldsymbol{y}) = \int_{\boldsymbol{x} \leq \boldsymbol{y}} t_\nu(\boldsymbol{x}; \boldsymbol{\mu}, \Sigma) d\boldsymbol{x} , \tag{5.7}$$

where the statement $\boldsymbol{x} \leq \boldsymbol{y}$ is understood as a component–wise comparison. This CDF does not have a closed–form solution but can be approximated through numerical integration. The t–copula is defined as the multivariate cumulative distribution given by

$$C_\nu^t(u_1, u_2, \ldots, u_n) = T_\nu\left(\left(t_\nu^{-1}(u_1), t_\nu^{-1}(u_2), \ldots, t_\nu^{-1}(u_n)\right); \boldsymbol{\mu}, \Sigma\right), \tag{5.8}$$

where $t_\nu^{-1}(u)$ is the inverse of the Student's t–distribution (5.4).

5.1.3 Generation of Dynamic Returns

The schematic depicted in Fig. 5.1 illustrates the integration of a dynamic optimization module, which transforms a series of historical returns into a dynamically modeled set of returns. These modeled returns are then fed into the portfolio optimization process. As mentioned in the introductory remarks of this section, the dynamic module aims to enhance the statistical robustness of the return sample for the optimization phase. This objective is accomplished through the following sequence of steps.

S1 Let $\{r_{i,t-T+1}, \ldots, r_{i,t}\}$ represent the historical return series for asset $i = 1, \ldots, n$ within the window $[t - T + 1, t]$. Fit[8] an ARMA(1,1)–GARCH(1,1)–Student's t innovation model to the return time series of each asset. This analysis will yield the parameters $\phi_{0,i}$, $\phi_{1,i}$, $\theta_{1,i}$, $\alpha_{0,i}$, $\alpha_{1,i}$, $\beta_{1,i}$, and ν_i for asset i, $i = 1, \ldots, n$.

S2 For asset $i = 1, \ldots, n$, from (5.3a) recursively calculate the shock series $\{a_{i,t-T+1}, \ldots, a_{i,t}\}$ using historical return values $r_{i,t}$ (with $a_{i,t-T} = r_{i,t-T} = 0$).

S3 For asset $i = 1, \ldots, n$, from (5.3c) and the shocks from S2, recursively calculate the variances $\{\sigma_{i,t-T+1}^2, \ldots, \sigma_{i,t}^2\}$ (with $\sigma_{i,t-T}^2 = 0$).

S4 For asset $i = 1, \ldots, n$, from (5.3b) and the results from S2 and S3, compute the innovation series $\{\epsilon_{i,t-T+1}, \ldots, \epsilon_{i,t}\}$.

6 In practice, Σ and $\boldsymbol{\mu}$ are estimated from sample data. Consequently, fitting the distribution $t_\nu(\boldsymbol{x}; \boldsymbol{\mu}, \Sigma)$ to a sample of multivariate data involves finding the best–fit value for ν.

7 Due to the symmetry of the distribution, $\boldsymbol{\mu}$ is also the median and mode of the distribution.

8 Fitting is done via a maximum likelihood estimation (MLE). Without going into further detail, MLE determines the values of the parameters which maximize the joint probability of the observed return data.

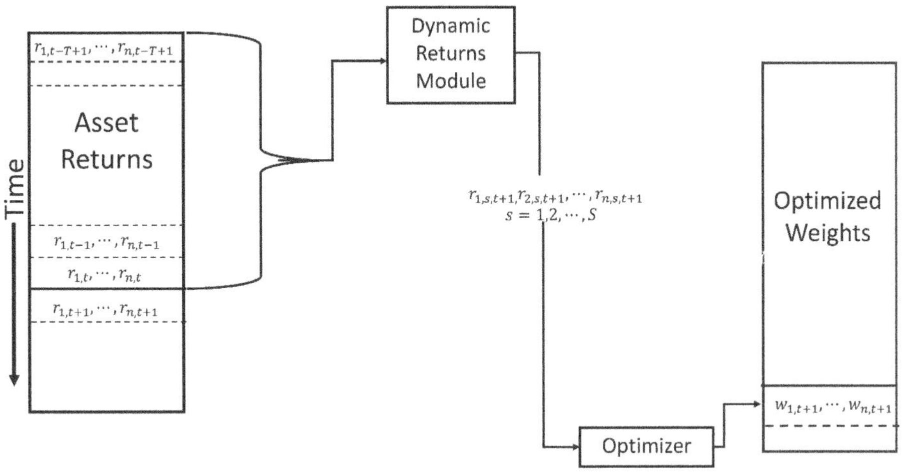

Fig. 5.1: Schematic of dynamic portfolio optimization

S5 Compute the copula transformations $u_{i,k} = T_{\nu_i}(\epsilon_{i,k})$, $k = t - T + 1, ..., t$, $i = 1, ..., n$, where T_{ν_i} is the CDF of the univariate Student's t–distribution, (5.5) for asset i.

S6 Estimate the parameters of the t–copula (5.8) for the set of transformed innovations $\{u_{i,t-T+1}, \ldots, u_{i,t}\}$, $i = 1, ..., n$. This fitting process employs an objective function designed to approximate the log–likelihood associated with the degrees of freedom parameter ν of the multivariate t–distribution.

S7 Produce $S \times n$ correlated samples, $\{u_{i,1}, \ldots, u_{i,S}\}$, $i = 1, ..., n$, by sampling from the t–copula. Note that S is significantly larger than T (the length of the historical window).

S8 Perform the inverse transformations $\epsilon_{i,s} = T_{\nu_i}^{-1}(u_{i,s})$, $s = 1, ..., S$, $i = 1, ..., n$.

S9 Using these transformed, correlated innovations with the parameters from S1, the shocks from S2, and variances from S3, construct a dynamic ensemble $\{r_{i,s,t+1}, s = 1, ..., S, i = 1, ..., n\}$ of return scenarios for $t + 1$ using (5.3),

$$\begin{aligned}
\sigma_{i,t+1}^2 &= \alpha_0 + \alpha_1 a_{i,t}^2 + \beta_1 \sigma_{i,t}^2, \\
a_{i,s,t+1} &= \sigma_{i,t+1} \epsilon_{i,s}, \\
r_{i,s,t+1} &= \phi_0 + \phi_1 r_{i,t} + a_{i,s,t+1} + \theta_1 a_{i,t}.
\end{aligned} \tag{5.9}$$

The collection of returns $\{r_{i,s,t+1};\ s = 1, \ldots, S,\ i = 1, \ldots, n\}$ constitutes the output from the dynamic module, which is subsequently supplied to the portfolio optimizer, as depicted in Fig. 5.1. CVaR–based optimization necessitates the full ensemble of return values, $\{r_{i,s,t+1};\ s = 1, \ldots, S,\ i = 1, \ldots, n\}$, while mean–variance optimization requires only the mean returns for each asset and the covariance matrix derived from this ensemble. The objectives of the dynamic formulation outlined at the beginning of this section are thereby achieved as follows. Fitting the ARMA(1,1)–GARCH(1,1)–Student's t innovation model in S1 enables the subsequent extraction (S2–S4) of a set of zero–mean, unit–variance innovations characterizing the marginal return distributions for each asset. Since tail information is precisely what is underrepresented in the historical window dataset, the transition to copula space (S5–S7) facilitates improved sampling of the tails of the correlated innovation distribution. Transformation of these samples under inverse marginal distributions (S8) enables computation of projected, correlated returns (S9) for time $t+1$. The net result of all these steps is the transformation of a $T \times n$ sample of returns into a (much) larger $S \times n$ sample that preserves statistical properties.

5.2 Empirical Examples: Dynamic Portfolio Optimization

We analyze the influence of a large sample of dynamic returns on the six optimizations of the cryptocurrency portfolio explored in Chapter 4. We utilize the same sized, $T = 2,000$, moving window in the dynamic optimizations that was used in the historical optimizations. Note that dynamic optimization is not applicable to the EWP and EWBH benchmark portfolios. We consider the following strategies.

5.2.1 Long–Only Strategy

Fig. 5.2 plots the cumulative return of each of the six dynamic optimizations under a long–only strategy with no turnover constraint ($C_{\text{TO}} = \infty$). Plotted for reference is the cumulative return of the benchmark EWBH portfolio. The corresponding return plots for the historical optimizations are provided in Fig. 4.1. Compared to historical optimization, dynamic optimization produces noticeable cumulative return (equivalently, price) improvement in the tangent portfolios. Compared to the historical optimization results, dynamic optimization shows some separation between the minimum–risk optimizations, particularly for MVP and M95, as well as some improvement in the cumulative return of the minimum–risk portfolios.

To gain some insight into these results, Fig. 5.3 shows the stacked bar graph of asset weights as a function of time for the MVP, M99, TVP and T99 portfolios. Also provided are plots of each asset's average weight over the time period. The analogous plots for historical optimization are provided in Fig. 4.2. Under historical optimization, the optimized compositions of MVP, M95 and M99 were dominated by BTC. Under dynamic optimization this dominance is lessened. Strong admixtures of other assets, noticeably ADA and QNT, as well as others were included. Similarly for TVP and T99, which were composed predominantly of SOL, SHIB and SAND with smaller admixtures of AVAX, AXS, QNT and NEAR under historic optimization.

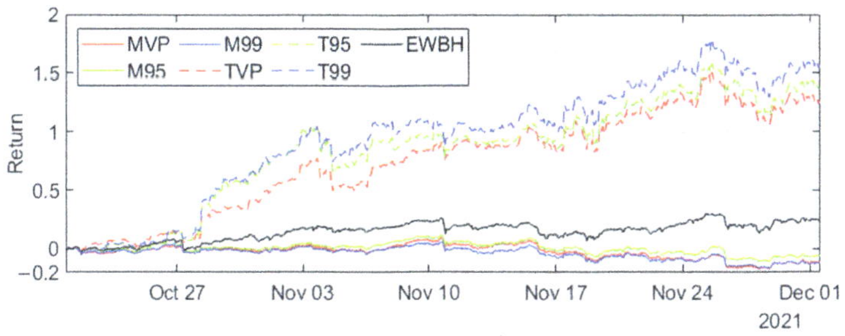

Fig. 5.2: Cumulative arithmetic return of the dynamic optimizations under the long-only strategy with no turnover constraint, and of the benchmark portfolio EWBH

As we have noted in Chapter 3, as one moves to higher expected returns on the mean–variance efficient frontier, the portfolio composition narrows, decreasing to a single asset at the largest possible expected return. In contrast, the minimum–variance portfolio will be the most diversified. (These observations are based strictly on the assumption that asset return distributions are Gaussian.) The plot of average weight over time provides insight into this behavior. MVP has a larger number of assets with non–negligible average weights than TVP. As CVaR$_\alpha$ for a Gaussian distribution is just a constant multiple of the standard deviation, the mean–CVaR portfolios should have the same efficient frontier behavior. However, T99 has a larger number of assets with higher average weights than does M99. This is due to the non–Gaussian nature of the tail regions of the cryptocurrency return distributions.

What stands out about dynamic optimization under no turnover constraint is the rate of change of portfolio composition (which leads to large turnover values). Fig. 5.4 provides box–whisker summaries of the distribution of turnover and L_2 norm values, as well as the bar graph of total turnover values, for the dynamic optimized portfolios of Fig. 5.2. The counterpart plots for the historical optimized portfolios are presented in Fig. 4.13. The dynamic optimization produces pronounced turnover, reflecting more frequent and larger asset weight changes resulting from increased sampling of, and sensitivity to changes in, the tail regions of the return distribution. Under dynamic optimization, with the exception of MVP, there are instances in which a total weight turnover value of 1 (100% turnover) occurs in a single rebalancing. For the tangent portfolios the Q_3 values exceed 90%, the Q_2 values are 70%, and the Q_1 values exceed 45%. For the minimum–risk portfolios, MVP has Q_3 and Q_1 values of 27% and 8% respectively; for M99 these values are 41% and 18%. Under historical optimization, the largest Q_3 value attained was 3.3% (by T99). Under dynamic simulation the total turnover for the tangent portfolios hovered around 670 (6.7×10^4% turnover), while M99 approached a total turnover value of 319. In contrast, under historical simulation T99 approached a total turnover of 33 while M99 had a total turnover of 0.017.

Given the large turnover values seen in Fig. 5.4, we investigated the imposition of the turnover constraint C_{TO}. We utilized the same set of values $C_{\mathrm{TO}} = \{0.04, 0.00134, 0.000114\}$

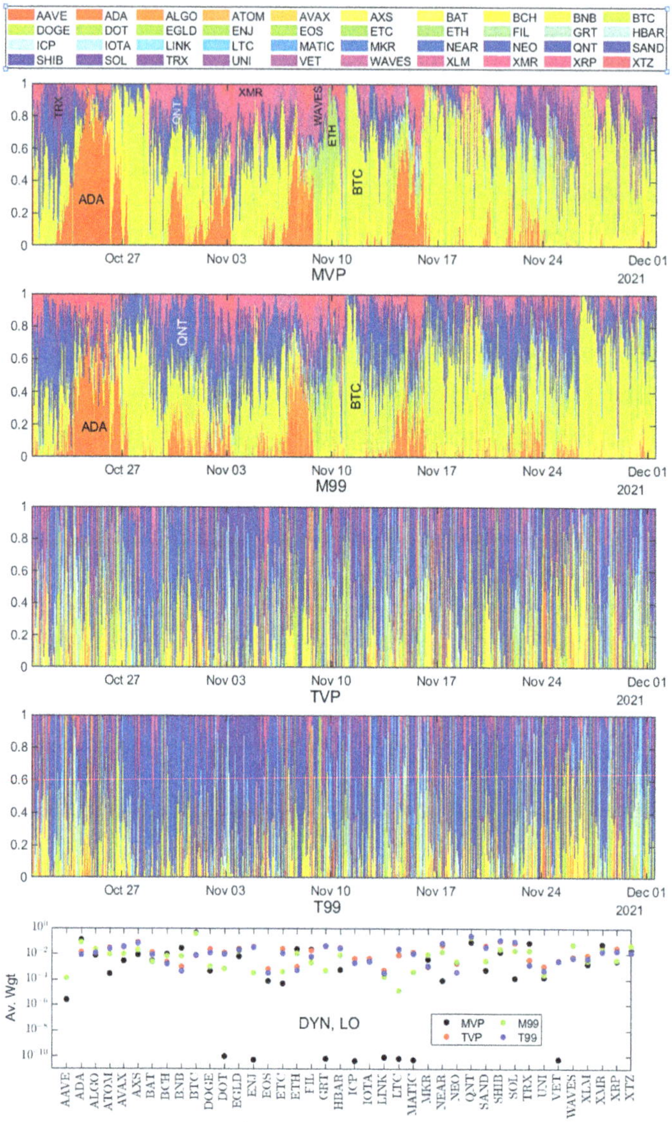

Fig. 5.3: Asset weights as a function of time for the dynamic optimizations MVP, TVP, M99 and T99 under the long-only strategy subject to no turnover constraint. (bottom) Time-averaged asset weights

used in Section 4.2. However, unlike historical historical optimization where a solution to the optimization problem was always obtained for any value of C_{TO} considered in Chapter 4, the optimization solvers[9] (quadprog for mean–variance and TrustRegionCP

9 Using the default solver parameters.

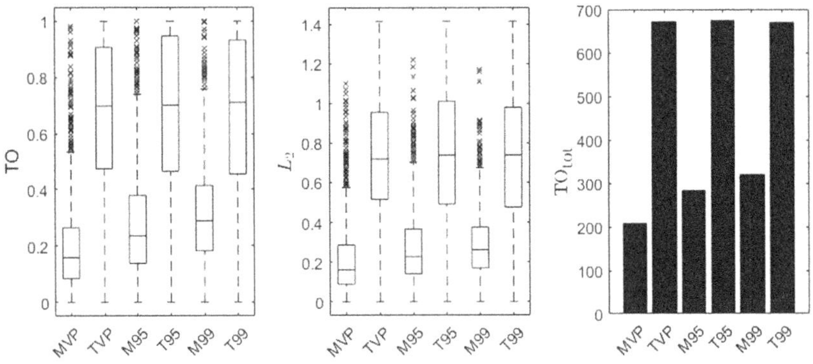

Fig. 5.4: Box-whisker summaries of the distribution of turnover and L_2 norm values, as well as the bar graph of total turnover values, for the dynamic optimizations subject to the long-only strategy with no turnover constraint

for mean–CVaR) failed to find a solution much more frequently under dynamic optimization. To handle this, when necessary we imposed a relaxation scheme on the value of C_{TO}. For example, for $C_{TO} = 0.04$, if the optimization to find the weights to apply to the trading interval $(t-1, t]$ failed to find a solution, the optimization was rerun using the relaxed value of $C_{TO} = 0.08$. If that optimization succeeded, the solution from the relaxed C_{TO} value defined the weights for the trading interval $(t-1, t]$. If the $C_{TO} = 0.08$ optimization failed, the optimization was rerun with $C_{TO} = 0.16$, with the obvious success/failure choices. The full sequence of relaxed values used was $C_{TO} = \{0.04, 0.08, 0.16, 0.32, \infty\}$, where $C_{TO} = \infty$ corresponds to a final attempt using no turnover constraint. In the very infrequent case where this last relaxation attempt value failed, the weights for the interval $(t-1, t]$ were set equal to the weights determined for the interval $(t-2, t-1]$. Regardless of whether, and how much of, the sequence was used for the time interval $(t-1, t]$, the optimization for the next time interval $(t, t+1]$ started with $C_{TO} = 0.04$. For the more stringent constraints $C_{TO} = 0.00134$ and 0.000114, we adopted more lenient (rather than lengthier) relaxation schedules. The specific relaxation schedules used were $C_{TO} = \{0.00134, 0.0134, 0.134, 0.27, \infty\}$ and $C_{TO} = \{0.000114, 0.00114, 0.0114, 0.114, 0.23, \infty\}$.

Tab. 5.1 shows the resultant success rates observed for these dynamic optimized, long–only portfolios under the different relaxation sequences. To understand the table, consider the M95 optimization run under the constraint $C_{TO} = 0.00134$. The optimization found a weight solution under this constraint value for 77.2% of the moving windows. When the constraint was relaxed to $C_{TO} = 0.0134$, solutions were obtained for an additional 3.2% of the moving windows. Relaxation to $C_{TO} = 0.134$ resulted in solutions for a further 14.5% of the windows, while relaxing to $C_{TO} = 0.27$ enabled solution for a further 3.0%. The remaining 1.7% of the moving windows required relaxing to $C_{TO} = \infty$. A solu-

Tab. 5.1: Success rates (in percent) for the different relaxation sequences for the dynamic optimizations under the long-only strategy

Sequence	MVP	TVP	M95	T95	M99	T99
∞	100.0	100.0	99.9	99.9	99.9	99.9
failed			0.1	0.1	0.1	0.1
0.04	100.0	99.5	83.0	89.0	82.6	88.4
0.08		0.5	6.0	1.9	5.7	1.9
0.16			5.0	4.0	5.9	3.9
0.32			3.7	2.9	3.7	3.9
∞			2.2	2.1	2.0	1.8
failed			0.1	0.1	0.1	0.1
0.00134	100.0	98.1	77.2	85.1	77.4	84.9
0.0134		1.7	3.2	2.2	3.5	1.3
0.134		0.2	14.5	7.6	13.8	8.4
0.27			3.0	3.4	3.5	3.8
∞			2.0	1.6	1.7	1.5
failed			0.1	0.1	0.1	0.1
0.000114	100.0	97.7	77.6	84.0	78.1	84.7
0.00114		1.0	0.4	0.3	0.4	0.2
0.0114		1.2	2.5	1.4	1.9	1.3
0.114		0.1	13.6	8.3	13.7	7.8
0.23			3.2		3.4	3.5
∞			2.6	5.9	2.4	2.4
failed			0.1	0.1	0.1	0.1

tion was obtained for all but one of these windows. Thus of the 1,000 moving windows in the out–of–sample data set, this relaxation schedule found solutions for 999 (99.9%).

As success (or failure) under a given value of C_{TO} in any specific relaxation sequence is independent of previous values attempted in that sequence, the entries in Tab. 5.1 can be reordered globally by C_{TO} to present cumulative success rates as a function of C_{TO}. These are presented in Tab. 5.2. For example, for M95 we see that 78% of the moving windows had a weight solution under the turnover constraint value $C_{TO} = 0.00114$, while 94% had a solution under the value $C_{TO} = 0.16$. While a non–decreasing value in the cumulative percentage is expected as the value of C_{TO} increases, Tab. 5.2 indicates that this was not always so. As the weight solution obtained for the time interval $(t-1, t]$ provides an initial guess for the weight solution for the interval $(t, t+1]$, the value of C_{TO} used for the solution on $(t-1, t]$ can have a subtle influence on the value of C_{TO} required for the solution for the interval $(t, t+1]$.

It is more instructive to view the data in Tab. 5.2 graphically. Fig. 5.5 plots the cumulative success rate as a function of C_{TO}. There is a clear separation between the mean–variance and mean–CVaR optimizations, with success rates much higher for

Tab. 5.2: Cumulative success rates (in percent) for the different values of C_{TO} from Tab. 5.1 (Omitted entries indicate 100%)

C_{TO}	MVP	TVP	M95	T95	M99	T99
0.000114	100.0	97.7	77.6	84.0	78.1	84.7
0.00114		98.7	78.0	84.3	78.5	84.9
0.00134		98.1	77.2	85.1	77.4	84.9
0.0114		99.9	80.5	85.7	80.4	86.2
0.0134		99.8	80.4	87.3	80.9	86.2
0.04		99.5	83.0	89.0	82.6	88.4
0.08		100.0	89.0	90.9	88.3	90.3
0.114			94.1	94.0	94.1	94.0
0.134			94.9	94.9	94.7	94.6
0.16			94.0	94.9	94.2	94.2
0.23			97.3	94.9	97.5	97.5
0.27			97.9	98.3	98.2	98.4
0.32			97.7	97.8	97.9	98.1
∞			99.9	99.9	99.9	99.9

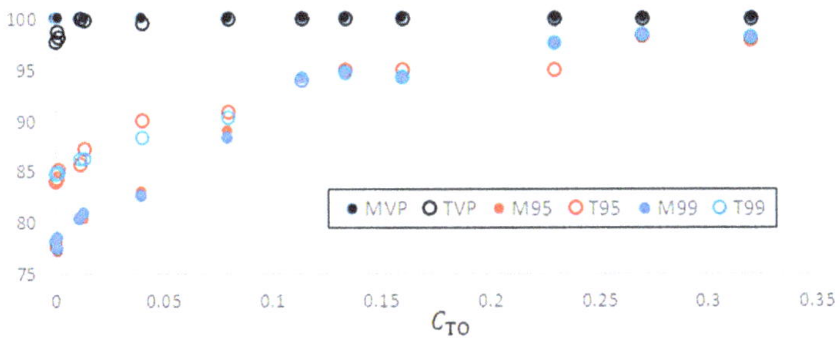

Fig. 5.5: Cumulative success rates (in percent) for the different values of C_{TO} from Tab. 5.2

mean–variance. There is also a clear separation between the mean–CVaR tangent and minimum–risk portfolios, with tangent portfolios having higher success rates for lower values of C_{TO}. By $C_{TO} = 0.114$, this difference between tangent and minimum–risk mean–CVaR portfolios essentially vanishes.

Fig. 5.6 presents the cumulative return for the dynamic optimized portfolios under the long–only strategy with $C_{TO} = 0.04$, 0.00134 and 0.000114. These can be compared to the results for no turnover constraint in Fig. 5.2. There was a marked drop in the cumulative returns of the tangent portfolios under imposition of the turnover constraint. For $C_{TO} = 0.04$, the cumulative returns T95 and T99 were comparable to that of the EWBH portfolio until approximately 20 November, when these returns fell below that of the EWBH. The cumulative return of the TVP remained comparable to the EWBH only until 2 November, after which it degraded to become comparable to the minimum–risk portfolios. Under the two more stringent values of C_{TO}, the cumulative returns of T95 and

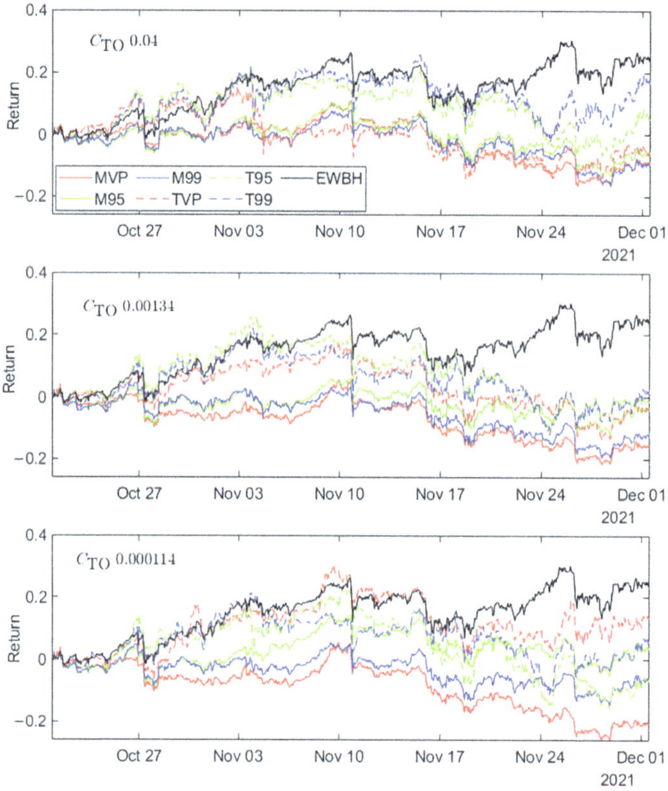

Fig. 5.6: Cumulative return for the dynamic optimizations under a long-only strategy with $C_{\mathrm{TO}} = 0.04$, 0.00134 and 0.000114. The cumulative return for the EWBH portfolio is included

T99 continued to degrade; in contrast, that of the TVP portfolio began to improve. As the value of C_{TO} decreased, the cumulative return of MVP degraded, while that of M99 and M95 improved, with greatest improvement for M95.

Fig. 5.7 shows the asset weight composition for the T99 portfolio as the turnover constraint C_{TO} value is decreased. These can be compared to the T99 results for no turnover constraint in Fig. 5.3. With no turnover constraint, the assets weights under went rapid changes. The sensitivity of CVaR_{99} dynamic optimization to changes in the negative tail region of the return distribution, combined with no turnover constraint, enabled the large cumulative returns seen in the tangent portfolios in Fig. 5.2. Imposition of a turnover constraint reduced the rate of such change and, consequently, reduced the cumulative return. Fig. 5.8 shows the average asset weight for the dynamic T99 portfolio under the four values of $C_{\mathrm{TO}} = \{\infty, 0.04, 0.00134, 0.000114\}$. The plots reveal that essentially the same assets were involved in forming the majority of the weight composition, even for the case $C_{\mathrm{TO}} = \infty$ where the asset weights changed rapidly.

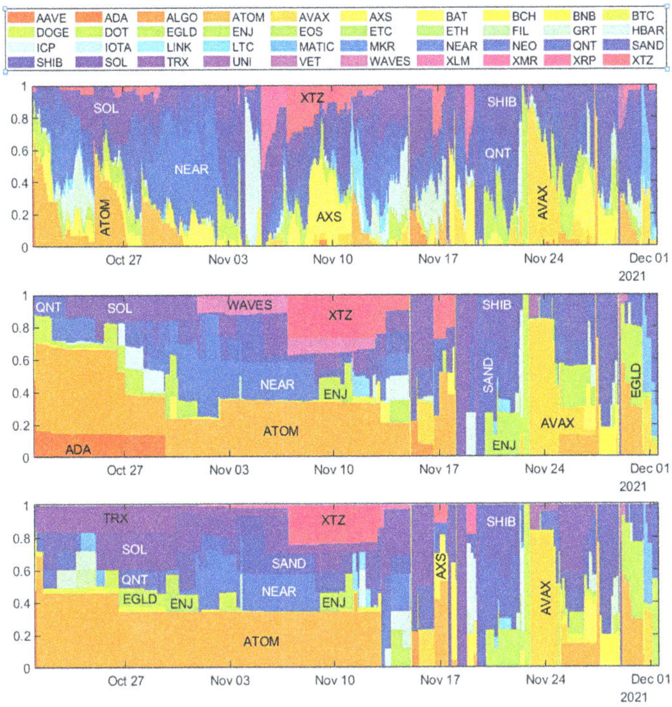

Fig. 5.7: Asset weights as a function of time for the dynamic T99 optimization under a long-only strategy with, from top to bottom, $C_{\mathrm{TO}} = 0.04$, 0.00134 and 0.000114

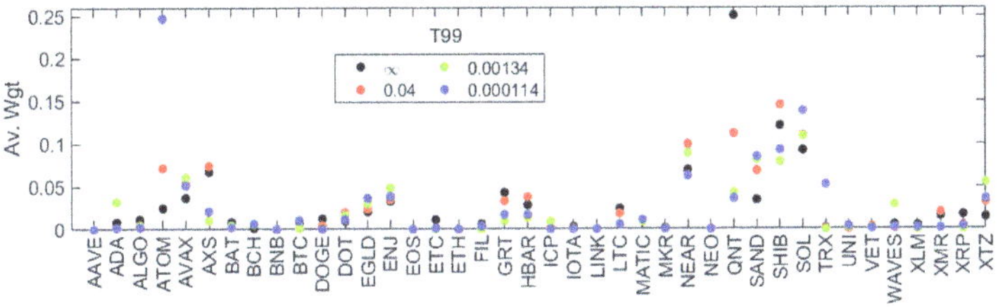

Fig. 5.8: Time-averaged asset weights for the dynamic T99 optimization under a long-only strategy with $C_{\mathrm{TO}} = \infty$, 0.04, 0.00134 and 0.000114

The historical optimized plots for T99 corresponding to those of Figs. 5.3 and 5.7 are shown in Fig. 4.15. While the assets SOL, SHIB, SAND and QNT dominated the T99 portfolio under historical optimization, a significant admixture of other assets (particularly ATOM) appeared under dynamic optimization, creating greater diversification. Unlike historical optimization, where decreasing the size of C_{TO} smoothed the stair-case effect seen in weight changes, under dynamic optimization, there were still periodic abrupt changes in assert weights. This is due to the relaxation sequence method used to handle a failure to find a weight solution. Under historical optimization with, for example, $C_{TO} = 0.00134$, there were no solution failures and every moving window weight solution was obtained under the constraint value $C_{TO} = 0.00134$. However, from Tab. 5.1, under dynamic optimization for T99 a solution attempt with $C_{TO} = 0.00134$ only succeeded 84.9% of the time. A further 8.4% of the solutions occurred under $C_{TO} = 0.134$, a relaxation by 100 in the value of C_{TO}. The result is the continued presence of abrupt changes in asset weights.

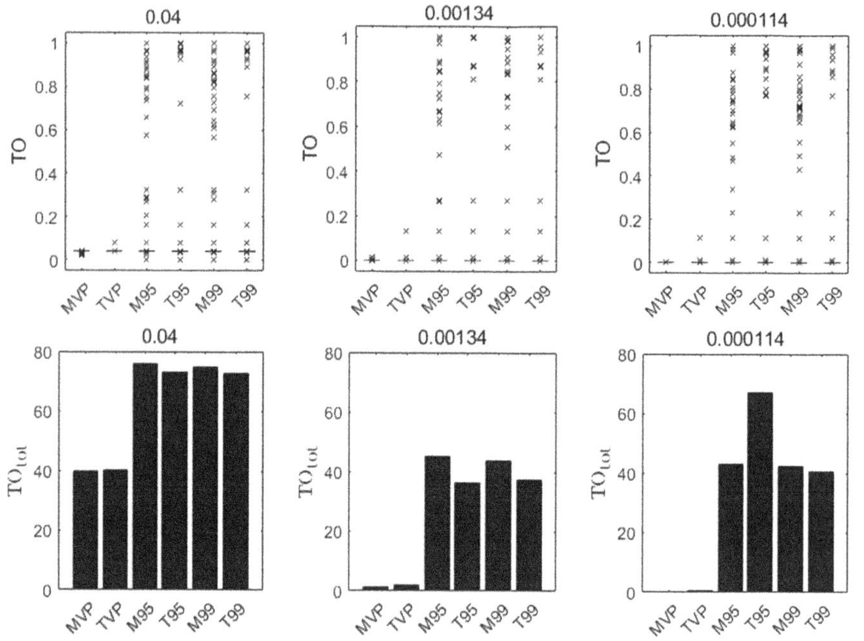

Fig. 5.9: (top) Box–whisker summary statistics of turnover TO and (bottom) total turnover TO_{tot} for the dynamic optimizations under a long–only strategy with $C_{TO} = 0.04$, 0.00134 and 0.000114

Fig. 5.9 displays the box–whisker summary statistics for the turnover TO and bar graphs of the total turnover TO_{tot} for the dynamic optimizations under the long–only strategy with $C_{TO} = \{0.04, 0.00134, 0.000114\}$. The plots for $C_{TO} = \infty$ are presented in Fig. 5.4.

The imposition of the $C_{TO} = 0.04$ constraint markedly reduced turnover of the dynamic optimizations, with further reduction evident under $C_{TO} = 0.00134$. However, imposition of $C_{TO} = 0.000114$ did not produce further reduction in turnover for the mean–CVaR portfolios, undoubtedly due to the fact that the solution success rate was essentially unchanged in changing from $C_{TO} = 0.00134$ to $C_{TO} = 0.000114$. From the box–whisker plots, it is also apparent that much of the increase in total turnover of the mean–CVaR optimizations compared to the mean–variance optimizations was due to larger outlier values.

The analogous plots for the historical optimizations are presented in Fig. 4.17. Comparison with the plots for dynamic optimizations shows the fact that turnover was higher for dynamic optimization at any value of C_{TO}.

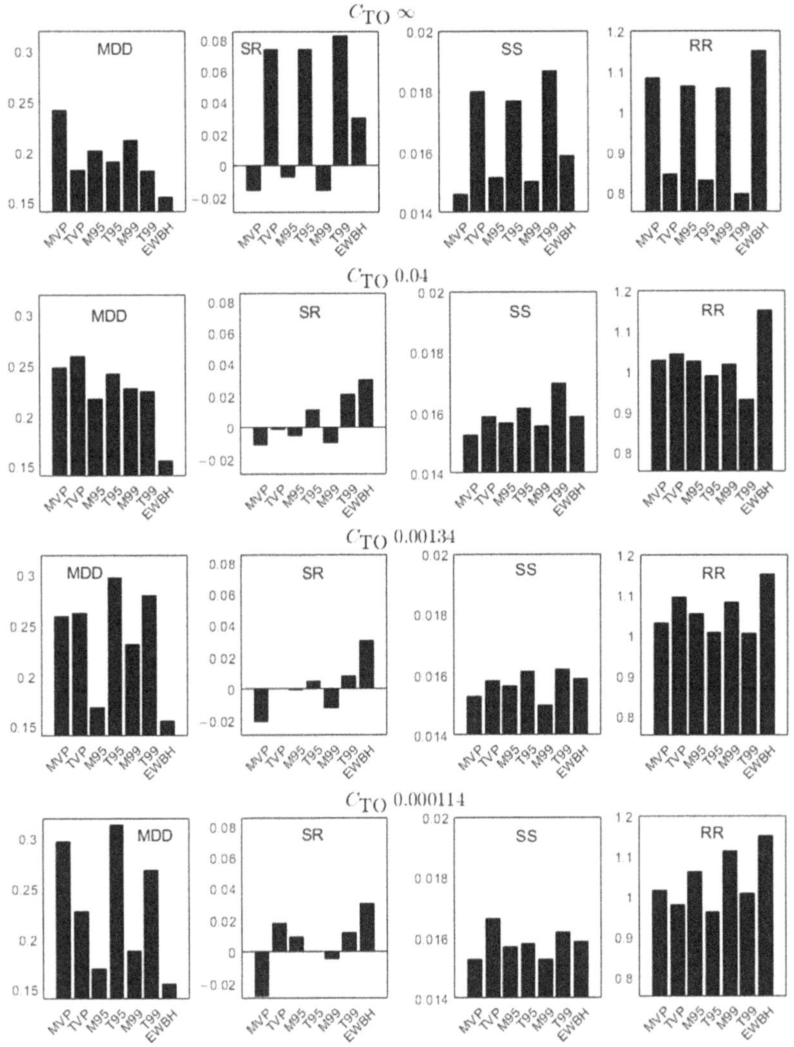

Fig. 5.10: Total risk measure values of the dynamic optimizations under the long–only strategy with $C_{\mathrm{TO}} = \infty$, 0.04, 0.00134 and 0.000114. Risk measures for the EWBH portfolio are included for comparison

Fig. 5.10 compares total risk measure values of the dynamic optimized portfolios under the long–only strategy with $C_{\mathrm{TO}} = \infty$, 0.04, 0.00134 and 0.000114. For comparison, the results for the EWBH benchmark portfolio are included. Relative to the benchmark

EWBH portfolio, the dynamic optimizations had worst MDD values; the MDD values for MVP, T95 and T99 became noticeably worse as the value of C_{TO} decreased. Under no turnover constraint, the Sharpe ratios of the tangent portfolios exceeded that of EWBH, however under the turnover constraints no dynamic portfolio outperformed EWBH. The total Sharpe ratios of the minimum–risk optimizations were almost always negative reflecting that, on average, their optimized hourly returns (which were larger than the risk–free rate) were negative over the time period. The tangent portfolios had better SS values than EWBH under no turnover constraint, and were able to maintain values competitive to that of the EWBH under the range of values of C_{TO} investigated. As under historical optimization (Fig. 4.22), the EWBH portfolio had the best RR. Compared to the historical optimizations, the dynamic minimum–risk portfolios were competitive in RR value with EWBH. As the value of C_{TO} increased, the value of RR also improved for the dynamic tangent portfolios. Compared with the historical optimization values of Fig. 4.22, generally speaking the dynamic MDD, SR and SS values were worse while the RR values were improved. Comparatively, the risk measures for the dynamic optimizations showed more variation under change of C_{TO} than did the risk measures for historic optimization.

The evaluation of optimization methods through return, turnover, and performance measures (MDD, SR, SS, and RR) provides a comprehensive, multidimensional perspective. However, this multidimensionality raises the question: how can we determine the best optimization method? This question does not yield a unique answer. From a statistical standpoint, various rank correlation tests[10] are available to assess whether two scoring systems produce correlated results (i.e., whether they are likely to generate similar rankings). Nevertheless, these tests do not resolve the fundamental question of which method performs best across multiple ranking systems.

Less sophisticated approaches to this problem involve deriving a final ranking for each method based on a weighted combination of its scores under each of a set of measures. Example approaches are:

- calculating the method's average score across the set of measures and ranking based on these averages;
- determining the median score for each method across the set of measures and ranking accordingly;
- ranking based on the frequency of first–place scores achieved by each method; and
- implementing an inverse ranking system based on the occurrence of last–place scores.

We employed the first approach to compare the performance of historical and dynamic optimizations for long–only cryptocurrency portfolios, examining both unconstrained

10 Common rank correlation tests include include Spearman's ρ, Kendall's τ, Goodman and Kruskal's γ, and Somer's D.

scenarios and those with a 4% hourly turnover constraint. We evaluate the six portfolio optimizations (MVP, M95, M99, TVP, T95, and T99) as well as the two benchmarks (EWP and EWBH) across five categories. These categories are:

C1 the price ratio PR = P(T)/P(0), where $P(T)$ represents the price at 13:00:00 on December 1, 2021 and $P(0)$ denotes the price at 21:00:00 on October 20, 2021; and
C2–C5 the aggregated values of MDD, SR, SS, and RR over the same time period.

Given the varying scales of these five measures, we normalize each category value to lie within the unit interval $[0, 1]$. To compare a set of n portfolios, for a given category measure $M_{\alpha,i}$, $\alpha = \{\text{PR, MDD, SR, SS, RR}\}$, corresponding to portfolio $i = 1, \dots, n$, the performance score $m_{\alpha,i}$ is

$$m_{\alpha,i} = \frac{M_{\alpha,i} - \min_{j=1,\dots,n} M_{\alpha,j}}{\max_{j=1,\dots,n} M_{\alpha,j} - \min_{j=1,\dots,n} M_{\alpha,j}}. \tag{5.10}$$

An average performance score for each portfolio is computed as

$$\widehat{m}_i = \frac{1}{5}\left[m_{\text{PR,i}} + (1 - m_{\text{MDD,i}}) + m_{\text{SR,i}} + m_{\text{SS,i}} + m_{\text{RR,i}}\right], \quad i = 1, \dots, n, \tag{5.11}$$

where the MDD performance scores are entered as $1 - m_{\text{MDD,i}}$ in the average to ensure that the portfolio with the lowest MDD receives the highest weighting.

Tab. 5.3 summarizes the final rankings based on the averaged performance scores for the eight portfolios. Under historical optimization, TVP always ranked first, while the other two tangent portfolios and EWBH always occupied ranks two through four, regardless of the value of C_{TO}. EWP, MVP, M99, and M95 occupied ranks five through eight, respectively, regardless of the value of C_{TO}.

Under dynamic optimization the behavior is much less regular. Under no turnover constraint, the three tangent portfolios occupy the top three ranks, while the minimum–risk portfolios occupy the bottom three ranks. As C_{TO} decreases:

- MVP continues to be lowest ranked;
- EWBH and EWP occupy ranks one and two respectively;
- the rankings of T99 and T95 fall;
- the rank TVP initially declines but then increases; and
- the ranking score of M95 increases.

Note that the scores of EWBH and EWP can change as the value of C_{TO} changes as, by (5.10), the values of $\max_j M_{\alpha,j}$ and $\min_j M_{\alpha,j}$ can change with C_{TO}.

Finally, we comment on the performance of the EWBH and EWP benchmark portfolios. They are employed for simplicity and because there really exist no appropriate benchmark portfolios for crypto assets comparable to, for example, the Dow Jones, the S&P 500, or the Russel 3000 indices used for equities. However, the EWP benchmark requires rebalancing every time step in order to maintain equal weights through time. It

Tab. 5.3: Average performance scores (5.11) of the benchmarks and the historical and dynamic optimized portfolios under the long-only strategy with $C_{TO} = \{\infty, 0.04, 0.00134, 0.000114\}$.

∞		0.04		0.00134		0.000114	
Historical							
TVP	0.657	TVP	0.674	TVP	0.687	TVP	0.810
T99	0.619	T95	0.635	T95	0.657	EWBH	0.702
EWBH	0.591	T99	0.600	T99	0.639	T95	0.650
T95	0.580	EWBH	0.594	EWBH	0.590	T99	0.565
EWP	0.397	EWP	0.411	EWP	0.395	EWP	0.523
MVP	0.072	MVP	0.087	MVP	0.060	MVP	0.134
M99	0.059	M99	0.073	M99	0.044	M99	0.132
M95	0.053	M95	0.069	M95	0.041	M95	0.125
Dynamic							
T99	0.741	EWBH	0.863	EWBH	0.936	EWBH	0.877
TVP	0.676	EWP	0.600	EWP	0.688	EWP	0.657
T95	0.649	T99	0.584	M95	0.494	TVP	0.635
EWBH	0.594	T95	0.392	TVP	0.443	M95	0.582
EWP	0.441	M95	0.261	T99	0.429	T99	0.496
M95	0.291	TVP	0.234	T95	0.374	M99	0.455
M99	0.232	M99	0.177	M99	0.244	T95	0.238
MVP	0.159	MVP	0.108	MVP	0.138	MVP	0.074

therefore incurs potentially tremendous transaction costs and is not practical to implement. In contrast, the EWBH portfolio undergoes no rebalancing, never incurs transaction costs, and is frequently implemented over relatively long periods of time. While it is theoretically susceptible to total investment loss, this passive investment strategy, employing a well-diversified set of assets, held long-term over a (long-term) healthy market economy, does surprisingly well. It can benefit from an initial (starting time) optimization rather than simple equal weighting (Hasan 2023). In contrast, see Fama and French (2010) for a critical analysis of active fund management.

5.2.2 Jacobs et al. Long-Short Strategy

We briefly turn our attention to the long-short strategies. We consider dynamic optimization under the Jacobs et al. strategy with the constraints $s = 0.1$ and $C_{TO} = 0.04$. Tab. 5.4 shows the success rate under the relaxation sequence for $C_{TO} = 0.04$. The mean-variance optimizations had the best solution success rates under the turnover restriction $C_{TO} = 0.04$, while the minimum-risk CVaR optimizations had the lowest. Interestingly, four of the 1,000 (0.4%) moving windows failed to find a solution under TVP optimization, even under no turnover constraint.

Tab. 5.4: Success rates (in percent) for the relaxation sequence of the dynamic optimizations under the Jacobs et. al strategy subject to the constraints $s = 0.1$ and $C_{TO} = 0.04$

Sequence	MVP	TVP	M95	T95	M99	T99
0.04	100.0	98.8	86.6	90.9	85.7	91.6
0.08		0.3	6.6	3.1	6.5	2.8
0.16		0.1	5.0	3.1	5.4	2.7
0.32		0.3	1.5	2.2	2.1	2.3
∞		0.1	0.3	0.7	0.3	0.6
failed		0.4				

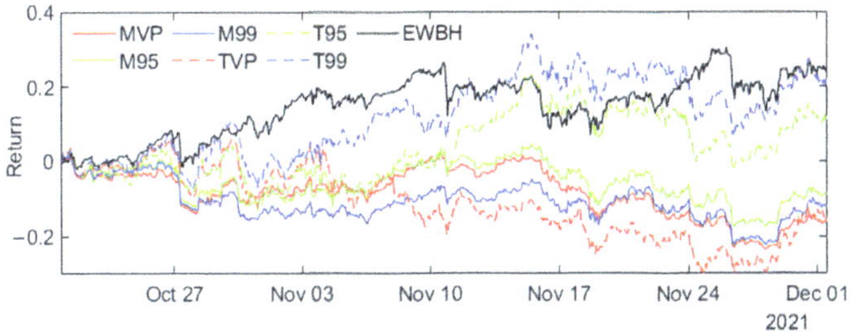

Fig. 5.11: Cumulative return for the dynamic optimizations under the Jacobs et al. strategy subject to the constraints $s = 0.1$ and $C_{TO} = 0.04$. The return for the EWBH benchmark is included for comparison

Fig. 5.11 plots the cumulative return of the dynamic optimized portfolios under this Jacobs et al. strategy. The benchmark portfolio EWBH, while long–only, is included in the plots as a test as to whether the "managed" dynamic portfolios outperformed a basic passive strategy over this time period when crypto assets were on a down slide. Compared to the long–only strategy with $C_{TO} = 0.04$ (Fig. 5.6), the cumulative return of the T99 and T95 portfolios improved. In contrast, the cumulative return of the remaining four optimized portfolios degraded.

Fig. 5.12 provides the asset weights as a function of time for the dynamic MVP, M99, TVP and T99 portfolios. As noted for the dynamic optimizations under the long–only strategy with turnover constraints, the use of relaxed values for C_{TO} to ensure a solution can produce discontinuous behavior in the asset composition. The T99 weight plot has asset labels added to show how the composition changes and how long the composition change can last. Notice that, even though M99 required even more frequent relaxation of C_{TO} than did T99, the M99 asset weights do not show the same discontinuous composition. This is one consequence of requiring a minimum–risk portfolio.

Note that the T99 asset weight plot only has positively weighted assets labeled. In spite of the appearance to the naked eye, the negatively weighted (shorted) assets appear in very narrow weight bands that are much more continuous throughout the entire time period, compared to the positive weight assets. This is the result of the $-s \le w_i < 0$

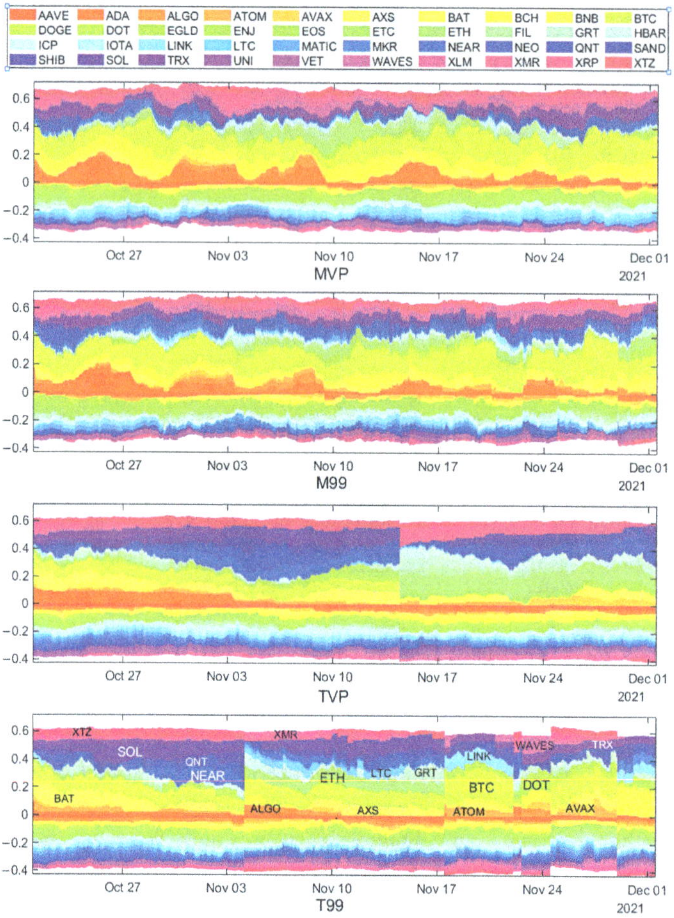

Fig. 5.12: Asset weights as a function of time for the dynamic MVP, M99, TVP and T99 optimizations under the Jacob et al. strategy subject to the constraints $s = 0.1$ and $C_{TO} = 0.04$

weight restriction applied to shorted assets, whereas positive weighted assets can have weights in the range $0 < w_i < 1 - s$.

The dynamic optimized T99 plot in Fig. 5.12 can be compared to the historic optimized T99 plot under the same long–short strategy constraints in Fig. 4.11. Greater diversification is evident under the dynamic optimization. This diversification is quantified in Fig. 5.13, which presents the average asset weight and fraction of hourly time periods an asset was shorted for these four dynamic optimized portfolios under this Jacobs et al. strategy. Under each optimization, virtually every asset is shorted for some percentage of the time. While BTC is the time–averaged dominant asset for MVP, M99 and T99, it is one of several equally dominant assets under TVP.

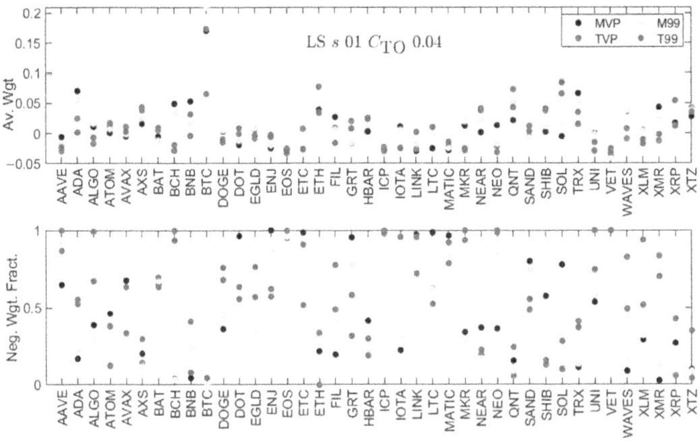

Fig. 5.13: (top) Average asset weight and (bottom) fraction of hourly time periods an asset was shorted for the indicated dynamic optimizations under the Jacobs et al. strategy with $s = 0.1$ and $C_{\mathrm{TO}} = 0.04$

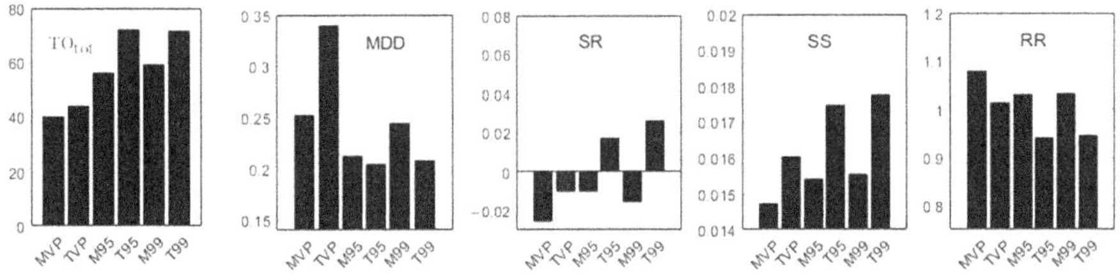

Fig. 5.14: Total turnover and risk measure values for the dynamic optimizations under the Jacobs et al. strategy with $s = 0.1$ and $C_{\mathrm{TO}} = 0.04$

The total turnover and total risk measures for all six optimized portfolios are presented in Fig. 5.14. Total turnover for the CVaR optimized portfolios improved compared to the long–only strategy of Fig. 5.9 under the same turnover constraint. Compared to the total risk measures for the long–only strategy for the same turnover constraint (Fig. 5.10), the most significant changes were an increase in the MDD for TVP, decreases in MDD for T95 and T99, improvements in SS for T95 and T99, and an improvement in RR for MVP. Compared to the total risk measures of Fig. 4.23 for the historical optimized portfolios under the Jacobs et al. strategy with $s = 0.1$ and $C_{\mathrm{TO}} = 0.04$, there was some improved of MDD for M95 and T95; and improvement of the RR values for the minimum–risk portfolios.

5.2.3 Lo–Patel Long–Short Strategy

For completeness, we consider the other long–short variant, the Lo–Patel strategy using a leverage constraint of $lev = 0.1$. We have noted earlier that the leverage acts as an effective turnover constraint. Therefore, under dynamic optimization, if a solution is not found under the constraint $lev = 0.1$, we adopt a relaxation schedule, employing the sequence $lev = \{0.1, 0.125, 0.15, 0.2, 0.3\}$. The success rates encountered for this relaxation sequence are provided in Tab. 5.5.

Tab. 5.5: Success rates (in percent) for the relaxation sequence of the dynamic optimizations under the Lo–Patel strategy subject to the constraint $lev = 0.1$

Sequence	MVP	TVP	M95	T95	M99	T99
0.1	100.0	96.4	94.8	90.1	93.3	90.4
0.125		0.4	1.7	1.7	3.1	0.9
0.15		0.2	1.0	2.1	0.9	2.0
0.2		0.1	1.4	2.0	1.0	1.5
0.3		0.2	0.7	1.2	1.2	1.8
failed		2.7	0.4	2.9	0.5	3.4

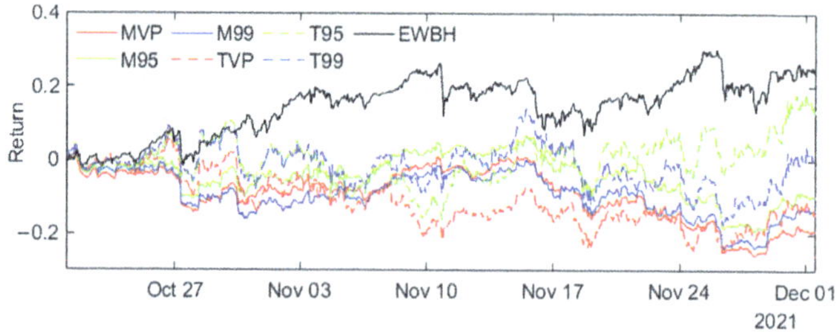

Fig. 5.15: Cumulative return for the dynamic optimizations under the Lo–Patel strategy subject to the constraint $lev = 0.1$. The return for the EWBH benchmark is included for comparison

The cumulative return plots are illustrated in Fig. 5.15. None of the dynamic optimized portfolios are competitive with the passive buy–and–hold strategy. To provide some insight into the behavior of the results of dynamic optimization under the Jacobs et. strategy of Section 5.2.2 and the Lo–Patel strategy of this section we concentrate on the two optimizations generating the highest cumulative returns, T95 and T99. We compare these against the T95 and T99 dynamic optimization under the long–only strategy with $C_{TO} = 0.04$ (Fig. 5.6). Both long–short optimizations suffer a significant drawdown during 30–31 October, 2021. Under the Jacobs et al. strategy, the T95 and T99 optimizations recover from this drawdown to be competitive with the EWBH portfolio by 13–15 November. Under the Lo–Patel strategy the T95 and T99 optimizations never recover (over the time period of the study) to be competitive with EWBH. In contrast, the long–only strategy suffers a drawdown over 29–31 October, from which it rapidly recovers to remain competitive with EWBH until 20 November, when both optimizations again suffer extended drawdowns.

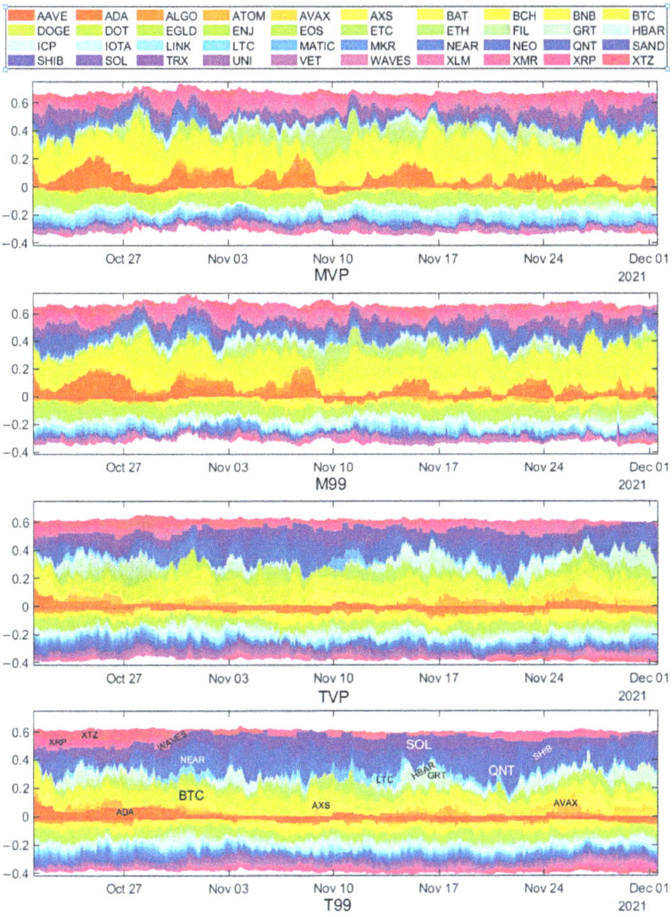

Fig. 5.16: Asset weights as a function of time for the dynamic MVP, M99, TVP and T99 optimizations under the Lo–Patel strategy subject to the constraint $lev = 0.1$

Asset weights as a function of time for the dynamic MVP, M99, TVP and T99 portfolios are provided in Fig. 5.16, while the average asset weight and the fraction of hourly time periods an asset was shorted are presented in Fig. 5.17. The results are qualitatively similar to those for the Jacobs et al. strategy of Section 5.2.2, with the exception that no abrupt changes in composition are seen. This undoubtedly stems from the fact that the relaxation sequence used for $lev = 0.1$ did not involve the large relative changes employed for the relaxation sequence of $C_{\mathrm{TO}} = 0.04$.

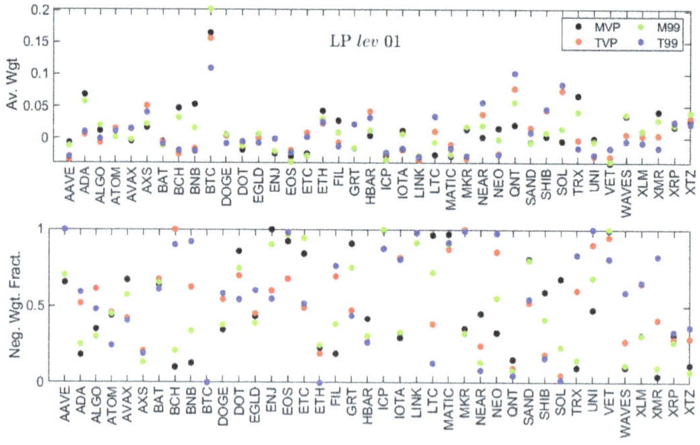

Fig. 5.17: (top) Average asset weight and (bottom) fraction of hourly time periods an asset was shorted for the indicated dynamic optimizations under the Lo–Patel strategy with *lev* = 0.1

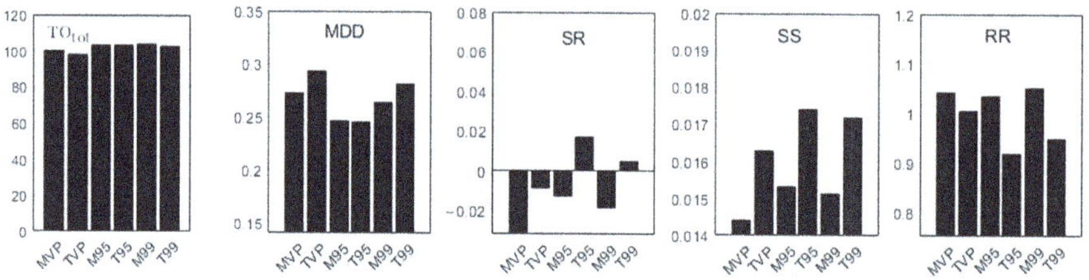

Fig. 5.18: Total turnover and risk measure values for the dynamic optimizations under the Lo–Patel strategy with *lev* = 0.1

Total turnover and risk measure values are provided in Fig. 5.18. Compared to the results in Section 5.2.2 for the Jacobs et al. strategy, total turnover was worse for Lo–Patel under all optimizations, with total turnover relatively independent of the optimization method. While MDD improved for TVP, it was worse for the other five optimizations. the total SR, SS and RR values showed relatively minor differences from those for Jacobs et al.

5.2.4 Closing Remarks – Dynamic Optimization

The behavior observed for the cryptocurrency portfolio under dynamic optimization is in contrast to what is observed in portfolios of equities (Lindquist et al. 2022, Chapter 5), indicating an unusual difference in crypto assets. Stocks are issued by companies whose performance is intensely scrutinized by equity analysts (both from the sell–side and the buy–side). There is no parallel for crypto assets. With the exception of stablecoins, crypto assets are essentially free–floating public ledgers. They are highly speculative digital assets driven by markedly different "fear and greed" behaviors compared to equities. This is one of the factors motivating the investigation of robust optimization techniques in Chapter 6.

It is possible that the ARMA(1,1)–GARCH(1,1)–t–innovation model investigated here for dynamic sample generation might require reworking for application to crypto assets. One possibility is that fractionally integrated ARMA–GARCH models are required. The fractionally integrated ARMA model, ARFIMA, allows for non-integer values $0 < d < 1$ of the differencing parameter used to produce stationary moving averages (Granger and Joyeaux 1980; Hosking 1981). (In Chapter 2.1 we showed stationarity of return series was obtained using the integer difference $d = 1$.) Similarly, the fractionally integrated GARCH model, FIGARCH, imposes a slower decay on the persistence of volatility shocks (Baillie et al. 1996).

6 Robust Portfolio Optimization

The methods proposed in Chapters 3–5 for constructing optimal portfolios rely on trusting the distribution of asset returns. In the case of historical optimization, this assumes that uncertainty is entirely captured by the empirical distribution. For dynamic optimization, the investor assumes that the time series model used to generate simulated returns for the next period accurately reflects the true (unobservable) multivariate distribution. This trust is overly optimistic in real markets, where various sources of information can alter return dynamics in ways not accounted for either in the historical returns or by the fitted model; particularly if the estimation window is kept relatively small to reflect recent market conditions and momentum. Motivated by the need to address uncertainty in the model specification used for portfolio optimization, a broad class of methodologies known as robust optimization (RO) has been developed.[1]

The term "robust" specifically refers to the goal of mitigating the negative effects of model misspecification by assuming that the model parameters, or certain other measures of the model, lie within a set of possible alternative values. For instance, one could assume that the estimated mean return for each asset i, $i = 1, \ldots, n$, (which may not be a direct model parameter, but is computed from the model parameters) derived from historical data should be more accurately represented by a confidence interval and not a single point value, with the width of this interval reflecting the degree of uncertainty associated with the estimate. A specialized approach of RO, known as distributionally robust optimization (DRO), treats the entire multivariate distribution of the random return vector $\boldsymbol{r}$ as uncertain and subject to model misspecification (Delage and Ye 2010).[2] Models belonging to this class formally characterize distributional uncertainty in terms of the distance between probability measures, looking at the model distribution used in traditional approaches as the center of an infinite set of alternative distributions within a predefined distance.

This chapter presents applications of RO and DRO to portfolio optimization, providing a general and intuitive overview of the methodology, its potential benefits, and its challenges. The goal is to introduce the reader to these broad class of models, which are mathematically more complex than those discussed in Chapters 3–5 but exhibit a similar level of numerical complexity during implementation. To keep the discussion accessible, we avoid mathematical proofs and complexities, directing interested readers to specialized sources. We first discuss RO models for mean–variance portfolios where the robustness is directed toward the mean of vector returns. The concept of a Kantorovich distance between random variables having countable outcomes is introduced as it is a necessary component in the application of the DRO model to mean–CVaR portfolios.

---

1 Ben-Tal et al. (2009) present a comprehensive introduction to RO.

2 See Lin et al. (2022) for a review of DRO theory and applications.

https://doi.org/10.1515/9781501517136-006

6.1 Uncertainty Sets for Expected Returns

In Section 3.1 we introduced mean–variance optimization, which relies on two sets of market information, the vector of mean returns of the assets and the covariance matrix of these returns. As in Section 3.1, let $\bar{\boldsymbol{r}} = [\bar{r}_1, \dots, \bar{r}_n]^\top$ and Σ denote, respectively, the true (unknown) mean and covariance of the assets.[3] From a practical point of view an investor needs to estimate these from market data; for example, directly as done in Chapters 3 and 4 or by fitting a time series model as in Chapter 5. We have assumed that, at a particular date, the investor observes asset returns $\boldsymbol{r}(t) = [r_1(t), r_2(t), \cdots, r_n(t)]^\top$ at historical times $t \in \{t_1, \dots, t_T\}$. We estimate the mean vector and the covariance matrix, respectively, through the point estimators

$$
\begin{aligned}
\hat{\boldsymbol{\mu}} = [\hat{\mu}_1, \dots, \hat{\mu}_n]^\top &= \frac{1}{T}\sum_{k=1}^{T} \boldsymbol{r}(t_k), \\
\hat{\Sigma}_{ij} &= \frac{1}{T-1}\sum_{k=1}^{T} (r_i(t_k) - \hat{\mu}_i)(r_j(t_k) - \hat{\mu}_j),
\end{aligned}
\tag{6.1}
$$

where $\hat{\mu}_i$ is a point estimate for the mean return $\bar{r}_i$ of asset i. In Section 3.1 we identify $\bar{\boldsymbol{r}}$ with $\hat{\boldsymbol{\mu}}$ and Σ with $\hat{\Sigma}$. Here, we maintain a distinction between the point estimates (6.1) and the true values.

Consider the alternate formulation (3.19) of the mean–variance problem (3.1),

$$
\max_{\boldsymbol{w}} \left(\beta \bar{\boldsymbol{r}}^\top \boldsymbol{w} - (1-\beta)\boldsymbol{w}^\top \Sigma \boldsymbol{w}\right), \quad \text{subject to } \boldsymbol{e}_n^\top \boldsymbol{w} = 1 . \tag{6.2}
$$

The parameter $\beta \in [0, 1]$ sets the strength of risk aversion. When $\beta = 1$, the solution of (6.2) is to invest all the wealth in the asset with the highest expected return, which is equivalent to the solution of problem (3.1) with the parameter r_p set to the maximum component of the mean vector $\bar{\boldsymbol{r}}$. As $\beta \downarrow 0$, the investor assigns a greater penalty (negative weight) to the portfolio variance, i.e. the investor expresses greater aversion to risk. $\beta = 0$ corresponds to the minimum variance solution MVP.

Since point estimates[4] are subject to estimation error, optimal solutions to (6.2) are affected by discrepancies between true and estimated parameter values, i.e., replacing $\bar{\boldsymbol{r}}$ and Σ in (6.2) with $\hat{\boldsymbol{\mu}}$ and $\hat{\Sigma}$ will change the optimal solution. The question is, how severely? We discuss RO and DRO by considering only a discrepancy in the mean value.

Assume that the true mean vector $\bar{\boldsymbol{r}}$ belongs to a set $\mathcal{U}$, the uncertainty set.[5] Then, for a fixed value of $\beta > 0$, the maximization must hold for the "smallest" value of $\bar{\boldsymbol{r}}$, that is for that member of the set $\mathcal{U}$ such that, for each fixed $\boldsymbol{w}$, $\bar{\boldsymbol{r}}^\top \boldsymbol{w}$ has minimum (worst

---

3 Theoretically, the return vector $\boldsymbol{r}$ is assumed to be defined on some probability space $(\Omega, \mathcal{F}, \mathbb{P})$. Complete knowledge of this space is required to know $\boldsymbol{r}$ and Σ.

4 Or indeed any type of estimate.

5 Equivalently, we regard $\mathcal{U}$ as a set of possible point estimates for $\bar{\boldsymbol{r}}$.

case) value. We can therefore reformulate the optimization problem (6.2) in terms of a Wald maximin problem (Wald 1945),

$$\max_{\boldsymbol{w}} \min_{\overline{\boldsymbol{r}}\in\mathcal{U}} \left[\beta \overline{\boldsymbol{r}}^\top \boldsymbol{w} - (1-\beta)\boldsymbol{w}^\top \Sigma \boldsymbol{w}\right] \quad \text{subject to } \boldsymbol{e}_n^\top \boldsymbol{w} = 1. \tag{6.3}$$

It should be emphasized that solving (6.3) is not equivalent to solving (6.2). Rather the solution to (6.3) gives a conservative "best worst utility" (i.e. a robust) value (Bertsimas and Sim 2004). The complexity of the above problem is determined by the complexity of the uncertainty set $\mathcal{U}$. In Sections 6.1.1 and 6.1.2, we present two forms of the uncertainty set, each of which produces a simplification in the formulation of problem (6.3).

6.1.1 Box Uncertainty Sets

The simplest way to define the uncertainty set is to establish lower and upper bounds for each asset's expected return. Specifically, we assume that the expected return of the i'th asset lies in the interval $\left[r_i^{(l)}, r_i^{(u)}\right]$, $i = 1,\dots,n$. The uncertainty set is $\mathcal{U} = \left\{\boldsymbol{r} \;\middle|\; r_i \in \left[r_i^{(l)}, r_i^{(u)}\right],\ i = 1,\dots,n\right\}$. A common approach to defining these bounds is to construct each interval symmetrically around the point estimate $\widehat{\mu}_i$ with radius δ_i, leading to the uncertainty set $\mathcal{U}_\delta(\widehat{\boldsymbol{\mu}}) = \{\boldsymbol{r} \mid |r_i - \widehat{\mu}_i| \le \delta_i,\ i = 1,\dots,n\}$. Under this assumption, the worst–case scenario for each asset depends on the sign of the portfolio weight w_i:

- if $w_i > 0$ (long position), the worst–case scenario is the lower bound $\widehat{\mu}_i - \delta_i$, producing the weighted expected return $r_i w_i = \widehat{\mu}_i w_i - \delta_i w_i$; and
- if $w_i < 0$ (short position), the worst–case scenario is the upper bound $\widehat{\mu}_i + \delta_i$, producing the weighted expected return $r_i w_i = \widehat{\mu}_i w_i - \delta_i |w_i|$.

Thus, for each asset the worst–case, weighted expected return can be expressed as $\widehat{\mu}_i w_i - \delta_i |w_i|$. The robust optimization (6.3) can be written

$$\max_{\boldsymbol{w}} \left[\beta \overline{\boldsymbol{r}}^\top \boldsymbol{w} - \beta \boldsymbol{\delta}^\top |\boldsymbol{w}| - (1-\beta)\boldsymbol{w}^\top \Sigma \boldsymbol{w}\right] \quad \text{subject to } \boldsymbol{e}_n^\top \boldsymbol{w} = 1, \tag{6.4}$$

where $\boldsymbol{\delta} = [\delta_1, \dots, \delta_n]^\top$ and $|\boldsymbol{w}| = [|w_1|, \dots, |w_n|]^\top$. As $|w_i| = \operatorname{sign}(w_i) w_i$, and $\delta_i > 0$, $i = 1, \dots, n$, we can write $\boldsymbol{\delta}^\top |\boldsymbol{w}| = \sum_{i=1}^n \operatorname{sign}(w_i)\delta_i w_i$, which explicitly formulates (6.4) as a linear optimization, solvable in polynomial time.

There are multiple ways to set the vector $\boldsymbol{\delta}$. As it quantifies the uncertainty of our point estimates $\widehat{\boldsymbol{\mu}}$, a straightforward approach is to use confidence intervals (CIs) to estimate $\boldsymbol{r}$. Assuming the central limit theorem holds – i.e., the asset returns are iid, have finite variance, and the sample size T is sufficiently large – we can use the standard normal CIs,

$$\widehat{\mu}_i \pm z_{\alpha/2} \frac{\widehat{\sigma}_i}{\sqrt{T}}, \qquad i = 1,\dots,n, \tag{6.5}$$

giving $\delta_{\alpha,i} = z_{\alpha/2}\widehat{\sigma}_i/\sqrt{T}$, where $\widehat{\sigma}_i = \sqrt{\widehat{\Sigma}_{ii}}$ is the sample standard deviation of the i'th asset, $z_{\alpha/2}$ is the critical value from the standard normal distribution, and the confidence level $1-\alpha$ determines the degree of robustness.

To enable graphical illustration, consider an example with $n = 2$ assets; specifically a portfolio composed of the two best–known crypto tokens, BTC and ETH. At 21:00:00 on October 20, 2021, the goal is to determine the uncertainty intervals $(\delta_{\alpha,\mathrm{BTC}}, \delta_{\alpha,\mathrm{ETH}})$ for the expected values of these two cryptocurrencies based on the last $T = 2000$, one–hour trades. Fig. 6.1 displays the uncertainty sets for three values of $\alpha = 0.2, 0.1, 0.05$. Also represented is the point estimate vector $\hat{\boldsymbol{\mu}} = [\hat{\mu}_{\mathrm{BTC}}, \hat{\mu}_{\mathrm{ETH}}]^\top$ for the mean return. For each choice of α, the corresponding lower and upper intervals define a rectangular region in the two–dimensional plane, where the axes represent the returns of the two assets. As α increases, the rectangular area expands, enlarging the uncertainty set.

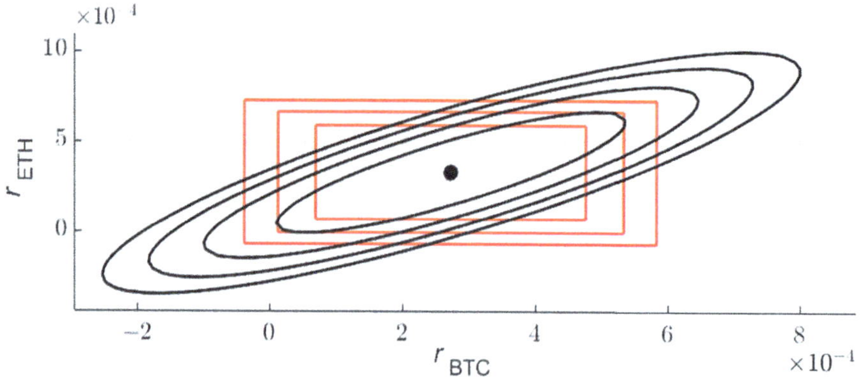

Fig. 6.1: Illustration of uncertainty sets for the mean vector of a portfolio consisting of BTC and ETH. Red lines represent different uncertainty boxes; black lines denote uncertainty ellipsoids. The black dot indicates the point estimate of the mean

Although simple, this example highlights that the box uncertainty sets are constructed independently for each asset, leading to the loss of information on their covariance structure. To address this limitation, ellipsoidal uncertainty sets have been introduced.

6.1.2 Ellipsoidal Uncertainty Sets

Let $\boldsymbol{Q}$ be a real, symmetric, positive definite matrix. Then $\boldsymbol{Q}^{-1}$ is also real, symmetric and positive definite. Consider the uncertainty set

$$\mathcal{U}_\gamma(\hat{\boldsymbol{\mu}}) = \left\{\boldsymbol{r} \;\middle|\; (\boldsymbol{r} - \hat{\boldsymbol{\mu}})^\top \boldsymbol{Q}^{-1}(\boldsymbol{r} - \hat{\boldsymbol{\mu}}) \le \gamma^2\right\}, \qquad \gamma \in \mathbb{R}. \tag{6.6}$$

The set $\mathcal{U}_\gamma(\hat{\boldsymbol{\mu}})$ is an ellipsoidal uncertainty set; its boundary,

$$(\boldsymbol{r} - \hat{\boldsymbol{\mu}})^\top \boldsymbol{Q}^{-1}(\boldsymbol{r} - \hat{\boldsymbol{\mu}}) = \gamma^2 \tag{6.7}$$

is a quadratic form which defines a hyperellipsoid (ellipsoid, for brevity) centered at the point estimate $\hat{\boldsymbol{\mu}}$. The (normalized) eigenvectors of $\boldsymbol{Q}^{-1}$ define the principal axes of the ellipsoid, while the inverse of each eigenvalue of Q^{-1} corresponds to the square of the length of the corresponding semi–axis. The matrices Q and Q^{-1} have the same eigenvectors; the eigenvalues of Q are the inverse of those for Q^{-1}. Thus, each eigenvalue of Q corresponds directly to the square of the length of the corresponding semi–axis. The left–hand side of (6.7) is the referred to as the ellipsoidal norm of the vector $r - \hat{\mu}$. The constant γ is analogous to the constant α for the box uncertainty set; γ linearly scales the ellipsoid uniformly in each direction, establishing the overall uncertainty of the set.

The optimization problem is now (6.3) with $\mathcal{U}$ replaced by $\mathcal{U}_\gamma$. We begin by solving the inner problem which involves minimizing the objective function subject to the constraint $(\overline{r} - \hat{\mu})^\top Q^{-1}(\overline{r} - \hat{\mu}) - \gamma^2 \le 0$. The Lagrangian function for this optimization problem is

$$\mathcal{L}(\overline{r}; \lambda) = \beta \overline{r}^\top w - (1 - \beta) w^\top \Sigma w + \lambda \left[(\overline{r} - \hat{\mu})^\top Q^{-1} (\overline{r} - \hat{\mu}) - \gamma^2 \right], \tag{6.8}$$

with λ being the Lagrange multiplier associated with the $\mathcal{U}_\gamma$ constraint. Requiring the gradient of (6.8) with respect to $\overline{r}$ to be zero (first–order optimality condition),

$$\nabla_{\overline{r}} \mathcal{L} = \beta w + \lambda 2 Q^{-1} (\overline{r} - \hat{\mu}) = 0, \tag{6.9}$$

produces the solution $\overline{r} = \hat{\mu} - \beta Q w / (2\lambda)$. The value of λ is then found by substituting the solution for $\overline{r}$ into (6.7), which leads to the solution $\lambda = \beta \sqrt{w^\top Q w} / (2\gamma)$. Substituting λ back into the solution for $\overline{r}$ gives

$$\overline{r} = \hat{\mu} - \gamma \frac{Q w}{\sqrt{w^\top Q w}},$$

which is the value for $\overline{r}$ that minimizes $\overline{r}^\top w$ subject to the elliptical uncertainty set constraint. The minimized value is

$$\overline{r}^\top w = \left[\hat{\mu} - \gamma \frac{Q w}{\sqrt{w^\top Q w}} \right]^\top w = \hat{\mu}^\top w - \gamma \sqrt{w^\top Q w}. \tag{6.10}$$

With this inner minimization problem solved, the outer maximization problem in (6.3) becomes

$$\max_{w} \left[\beta \hat{\mu}^\top w - \beta \gamma \sqrt{w^\top Q w} - (1 - \beta) w^\top \Sigma w \right] \quad \text{subject to } e_n^\top w = 1. \tag{6.11}$$

We assume the historical observations $r_i(t_k)$, $k = 1, \ldots, T$, $i = 1, \ldots, n$, are iid samples drawn from a multivariate t–distribution with $\nu > n - 1$ degrees of freedom. The point estimate for the mean $\hat{\mu}$ and the sample covariance matrix Σ are given by (6.1).

By replacing the general matrix Q with the covariance matrix Σ, the ellipsoidal norm now represents the squared Mahalanobis distance between the vectors $\overline{r}$ and $\hat{\mu}$. As noted, we assume the historical observations are drawn from a multivariate t–distribution with $\nu > n - 1$ degrees of freedom. For T sufficiently large, under the multivariate central limit theorem the t–distribution approaches the normal distribution.

Thus the $100(1-\alpha)\%$ confidence region for the mean vector is well approximated by the ellipsoid

$$(\bar{\boldsymbol{r}} - \hat{\boldsymbol{\mu}})^\top \hat{\Sigma}^{-1} (\bar{\boldsymbol{r}} - \hat{\boldsymbol{\mu}}) \le \chi^2_{n,\alpha}\,, \tag{6.12}$$

where $\chi^2_{n,\alpha}$ is the α critical value[6] of the χ^2–distribution. Using the $n = 2$ portfolio example from Section 6.1.1, elliptic uncertainty sets are depicted in Fig. 6.1 for values of α = 0.0014, 0.0028, 0.0042 and 0.0055. Ellipsoidal uncertainty sets retain knowledge of the linear correlation structure embedded in the covariance matrix of the asset returns, in this case capturing the positive correlation between the 1–hour returns of BTC and ETH.

We conclude with the following observation. Cryptocurrency returns are typically not iid,[7] exhibiting features such as autocorrelation and heteroskedasticity. Potential improvements to the construction of CIs to account for such characteristics include: adjusted CIs using either Newey–West or heteroskedasticity–robust standard errors; bootstrap CIs, and Bayesian approaches (see, e.g., Hamilton 2020).

6.2 Kantorovich Distance between Discrete Random Variables

The distance between probability measures is one way to quantify the difference between two probability distributions. Similar to the concept of vector norms, it can be defined in several ways. We discuss the Kantorovich distance (KD) for discrete distributions, and demonstrate its application to portfolio optimization. This distance is related to the mass transportation problem; it is in this context that we introduce it.

The Kantorovich (1939) formulation of the mass transportation problem can be expressed as the following general problem. Let X and Y be two separable metric spaces such that any probability measure on X or Y is a Radon measure. Let μ and ν be probability measures on X and Y, respectively. Let $c : X \times Y \to [0, \infty)$ be a Borel–measurable function. Let $\Gamma(\mu, \nu)$ denote the set of all probability measures on $X \times Y$ having marginals μ on X and ν on Y. Find the probability measure $\gamma \in \Gamma$ that satisfies

$$\mathcal{T}_c(\mu, \nu) = \inf \left\{ \int\limits_{X\times Y} c(x, y) d\gamma(x, y) \mid \gamma \in \Gamma(\mu, \nu) \right\}. \tag{6.13}$$

The non–negative function $c(x, y)$ represents the cost to move a unit of measure from $x \in X$ to $y \in Y$ such that the total marginals μ and ν are retained.

The Kantorovich formulation was an improvement on a volume transportation problem formulated by Monge (1781) (of whose work Kantorovich was unaware). The Monge formulation involved the specification of a transport map $F : X \to Y$ and its "push forward" $F_*(\mu)$ defined as $F_*(\mu)(B) = \mu(F^{-1}(B))$ for any measurable set

6 The value of the quantile function $(\chi^2)^{-1}(1 - \alpha)$.

7 Nor, indeed, are the returns for the assets within and across all asset classes.

$B \in Y$. The Monge formulation can be ill–posed as there may not exist an F such that $F_*(\mu) = \nu$ for particular choices of μ and ν. The Kantorovich transportation problem was then further generalized to include multiple transshipment (transit) points (the Kantorovich–Rubinstein problem) or multiple stages (the Kantorovich–Rubinstein–Kemperman problem).[8] For a general cost function c, $\mathcal{T}_c$ is referred to as the Kantorovich functional.

Kantorovich considered the case in which the two metric spaces were identical, $X = Y = U$, with U characterized by the metric $d(x, y)$, $x, y \in U$. Hence μ and ν are two different probability measures on U. In the special case where the cost function is $c = d$, the Kantorovich functional $\mathcal{T}_c$ can be defined as the Kantorovich distance (KD)

$$\mathcal{T}_d(\mu, \nu) = \inf \left\{ \int_{U \times U} d(x, y) d\gamma(x, y) \mid \gamma \in \Gamma(\mu, \nu) \right\}. \tag{6.14}$$

The KD can be used as a measure of the distance between two probability measures. When $U = \mathbb{R}^n$, useful distance measures are $d(x, y) = ||x - y||^p$, $p \in [1, \infty)$. Restricting to $U = \mathbb{R}$, the KD can be written[9]

$$\mathcal{T}_p(\mu, \nu) = \left(\inf \left\{ \int_{\mathbb{R} \times \mathbb{R}} ||x - y||^p d\gamma(x, y) \mid \gamma \in \Gamma(\mu, \nu) \right\} \right)^{1/p}. \tag{6.15}$$

To make the discussion more intuitive, we consider the application of the mass transportation problem to discrete, real–valued random variables. Consider two discrete, real–valued, random variables, X and Y, with the same number T of possible outcomes.[10] The set of outcomes for X is a separable metric space, with the same statement for Y. Their probability measures are their probability mass distributions (pmfs) p_X and p_Y, which can be denoted

$$p_X = \{(x_i, p_i),\ i = 1, \ldots, T\}, \quad p_Y = \{(y_j, q_j),\ j = 1, \ldots, T\}. \tag{6.16}$$

As these are distributions, the total mass of each measure must sum to one; $\sum_{t=1}^{T} p_t = \sum_{t=1}^{T} q_t = 1$. Consider the problem of moving $\eta_{i,j}$ of the mass $p_X(x_i)$ to form $\eta_{i,j}$ of p_Y at y_j. We must have the restrictions on the elements $\eta_{i,j}$ of the $T \times T$ matrix $\boldsymbol{\eta}_{X,Y}$,

$$\eta_{i,j} \geq 0, \quad \sum_{i=1}^{T} \eta_{i,j} = p_X(x_i) = p_i, \quad \sum_{j=1}^{T} \eta_{i,j} = p_Y(y_j) = q_j. \tag{6.17}$$

8 For a complete presentation, see the work of Rachev (1985) and Rachev and Rüschendorf (1998a,b).

9 Apparently unaware of the work by Kantorovich, R. L. Dobrushin coined the phrase Wasserstein distance for the case of general p, attributing it to the work of Vaseršteĭn (1969) (commonly written in the German form Wasserstein). The term Wasserstein distance has spread through the literature; however, the general usage of Wasserstein distance rightfully should by attributed to Kantorovich (Deng and Du 2009).

10 The assumption of an equal number of outcomes is not necessary; it is introduced to simplify the notation.

Thus $\boldsymbol{\eta}_{XY}$ is some joint probability mass distribution having marginal pmfs p_X and p_Y (i.e., $\boldsymbol{\eta}_{XY} \in \Gamma(p_X, p_Y)$).[11] With $c(x_i, y_j) = d(x_i, y_j) = |x_i - y_j|$ (i.e. $p = 1$) representing the cost to transfer unit mass from x_i to y_j, this discrete mass transportation problem is

$$\mathcal{T}_1(p_X, p_Y) = \min_{\eta_{ij}} \sum_{i=1}^{T} \sum_{j=1}^{T} |x_i - y_j| \eta_{i,j}, \tag{6.18}$$

subject to (6.17).

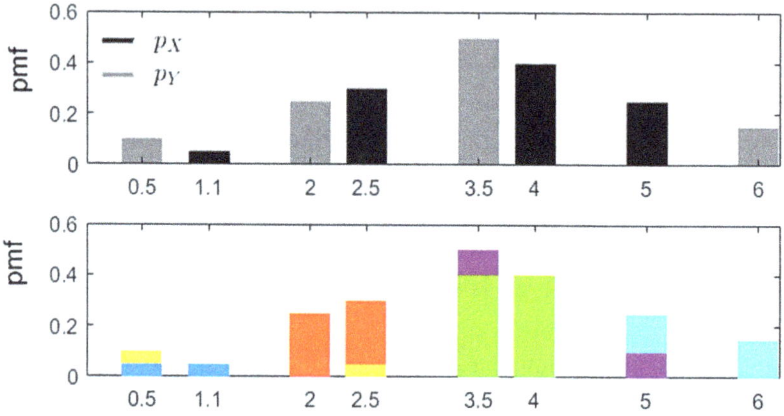

Fig. 6.2: Example of an optimal transport problem. The top panel illustrates the pmf of the source distribution X and that of the target distribution Y. The object is to transport p_X to p_Y under a plan that minimizes a total cost of all mass movements. The bottom panel visualizes the optimal transport solution; each color indicates the probability mass transported from p_X to construct p_Y.

Consider the following empirical example. The discrete random variables X and Y have the pmfs

$$p_X = \begin{cases} (1.1, 0.05), \\ (2.5, 0.30), \\ (4.0, 0.40), \\ (5.0, 0.25), \end{cases} \qquad p_Y = \begin{cases} (0.5, 0.10), \\ (2.0, 0.25), \\ (3.5, 0.50), \\ (6.0, 0.15). \end{cases}$$

These pmfs are illustrated in the top panel of Fig. 6.2. The object is to move p_X to produce p_Y in such a way that the total cost of all mass movement $\mathcal{T}_1$ is minimized. The optimal solution is

11 A simple example is the cross product distribution $\boldsymbol{\eta}_{XY} = p_X \times p_Y$. This however may not be the joint pmf needed to optimize (6.13).

$$\eta^*_{X,Y} = \begin{array}{c} \\ 1.1 \\ 2.5 \\ 4.0 \\ 5.0 \end{array} \begin{array}{cccc} 0.5 & 2.0 & 3.5 & 6.0 \\ \end{array} \begin{pmatrix} 0.05 & 0.0 & 0.0 & 0.0 \\ 0.05 & 0.25 & 0.0 & 0.0 \\ 0.0 & 0.0 & 4.0 & 0.0 \\ 0.0 & 0.0 & 0.1 & 0.15 \end{pmatrix}$$

with $\mathcal{T}_1(p_X, p_Y) = 0.655$. This solution is shown in the bottom panel of Fig. 6.2.[12]

Viewing the KD as a p–distance between two probability measures, we consider the probability measure p_X to be a discrete, empirical probability distribution obtained at time t from a set of historical return observations $\{\boldsymbol{r}(k), k = t - T + 1, ..., t\}$, $\boldsymbol{r}(k) = [r_1(k), ..., r_n(k)]^\top$ of n assets. We shall label such an empirical measure as $\widehat{p}_T$ to indicate that it arises from T observations. Rather than specify the measure p_Y, we formulate a distance–between–measure problem by considering the set

$$B_\epsilon(\widehat{p}_T) = \{p_Y : \mathcal{T}_p(p_Y, \widehat{p}_T) \le \epsilon\}. \tag{6.19}$$

This set contains distributions whose KD from $\widehat{p}_T$ is at most ϵ. (Note that there is no requirement that the distributions in the ball be either discrete or have T realizations. However, it does contain all T–sized discrete distributions fulfilling the distance restriction.) Equation (6.19) specifies a ball of uncertainty around the empirical distribution. Specifying higher values for ϵ increases the ambiguity, making the empirical measure less certain.

6.3 Robust CVaR Optimization on Historical Returns

We describe the DRO (Delage and Ye 2010; Lin et al. 2022) approach using mean–CVaR optimization with historical returns (Section 3.2).[13] In that section we considered the CVaR$_\alpha$–minimization problem (3.25). As per the discussion of (3.20), (3.25) can alternately be written

$$\min_{\boldsymbol{w}} \left\{-\beta\overline{\boldsymbol{r}}^\top \boldsymbol{w} + (1-\beta)\,\mathrm{CVaR}_\alpha(\boldsymbol{w})\right\}, \quad \text{subject to } \boldsymbol{e}_n^\top \boldsymbol{w} = 1\,. \tag{6.20}$$

---

12 It is fairly straightforward to confirm this must be the optimum solution. To maintain a minimal distance–based cost: $p_X(1.1, 0.05)$ must transport to $p_Y(0.5, 0.05)$; $p_X(4.0, 0.4)$ must transport to $p_Y(3.5, 0.4)$; $p_X(2.5, 0.25)$ must transport to $p_Y(2.0, 0.25)$; and $p_X(5.0, 0.15)$ must transport to $p_Y(6.0, 0.15)$. At this point the masses $p_X(1.1, 0.0)$ and $p_X(4.0, 0.0)$ are depleted, while the masses $p_Y(2.0, 0.25)$ and $p_Y(6.0, 0.15)$ are filled. There remains the problem of transporting the remaining $p_X(2.5, 0.05)$ and $p_X(5.0, 0.1)$ between $Y = 0.5$ and $Y = 3.5$. It is cheaper to move the larger mass amount $p_X(5.0, 0.1)$ a smaller distance and the smaller mass amount $p_X(2.5, 0.05)$ a longer distance. Hence $p_X(5.0, 0.1)$ must transport to $Y = 3.5$, filling the mass $p_Y(3.5, 0.4) + p_Y(3.5, 0.1) = p_Y(3.5, 0.5)$, while $p_X(2.5, 0.05)$ must transport to $Y = 0.5$, filling the mass $p_Y(0.5, 0.05) + p_Y(0.5, 0.05) = p_Y(3.5, 0.1)$.

13 This keeps the mathematics simple while explaining the main ingredients of DRO.

Recall that $\overline{\boldsymbol{r}}$ is the (time–independent)[14] vector of mean returns of each asset in the portfolio. The solution of the problem depends on the multivariate probability distribution $p(\boldsymbol{r}(t))$ that governs the time–dependent vector $\boldsymbol{r}(t)$ of asset returns. To indicate uncertainty in knowledge of $p(\boldsymbol{r}(t))$, following Esfahani and Kuhn (2018), DRO considers the optimization problem,

$$\min_{\boldsymbol{w}} \mathbb{E}_{p(\boldsymbol{r}(t))}\left[-\overline{\boldsymbol{r}}^{\mathsf{T}}\boldsymbol{w} + \beta\,\mathrm{CVaR}_{\alpha}(\boldsymbol{w})\right], \quad \text{subject to } \boldsymbol{e}_n^{\mathsf{T}}\boldsymbol{w} = 1, \tag{6.21}$$

where the notation $\mathbb{E}_{p(\boldsymbol{r}(t))}[\cdot]$ indicates the expectation is taken with respect to the distribution function $p(\boldsymbol{r}(t))$. Thus as $p(\boldsymbol{r}(t))$ changes, so will the mean vector $\overline{\boldsymbol{r}}$.

Application of the Esfahani and Kuhn approach to (6.20) for the discrete problem (3.30), consisting of a historical window of asset return vectors $\boldsymbol{r}(t)$, $t = 1, \ldots, T$, and using the fact that the value of $\boldsymbol{w}$ that minimizes $\mathrm{CVaR}_{\alpha}(\boldsymbol{w})$ also minimizes $\mathrm{VaR}_{\alpha}(\boldsymbol{w})$, leads to

$$\begin{aligned} &\min_{\boldsymbol{w},\gamma} \mathbb{E}_{p(\boldsymbol{r}(t))}\left[-\beta\overline{\boldsymbol{r}}^{\mathsf{T}}\boldsymbol{w} + (1-\beta)\left(\gamma + \frac{1}{\alpha}\left(-\overline{\boldsymbol{r}}^{\mathsf{T}}\boldsymbol{w} - \gamma\right)^{+}\right)\right], \\ &\quad \text{subject to } \boldsymbol{e}_n^{\mathsf{T}}\boldsymbol{w} = 1. \end{aligned} \tag{6.22}$$

In contrast to (3.30), (6.22) allows for variability in the multivariate probability distribution $p(\boldsymbol{r}(t))$. Let $\widehat{p}_T := p(\boldsymbol{r}(t))$ denote the probability distribution obtained from the historical window. The data–driven DRO reformulation of problem (6.20) can be written as:

$$\begin{aligned} &\min_{\boldsymbol{w},\gamma} \max_{q\in B_{\epsilon}(\widehat{p}_T)} \mathbb{E}_q\left[-\beta\overline{\boldsymbol{r}}^{\mathsf{T}}\boldsymbol{w} + (1-\beta)\left(\gamma + \frac{1}{\alpha}\left(-\overline{\boldsymbol{r}}^{\mathsf{T}}\boldsymbol{w} - \gamma\right)^{+}\right)\right], \\ &\quad \text{subject to } \boldsymbol{e}_n^{\mathsf{T}}\boldsymbol{w} = 1. \end{aligned} \tag{6.23}$$

Each distribution $q(\boldsymbol{r}(t))$ defines a multivariate probability distribution for the return random vectors $\boldsymbol{r}(t)$, and hence for the average vector $\overline{\boldsymbol{r}}$. Because the minimization in (6.23) is taken using the worst–case scenario distribution function $q \in B_{\epsilon}(\widehat{p}_T)$, the efficient frontier determined by the solution of (6.21) with $p(\boldsymbol{r}(t)) = \widehat{p}_T$ and the efficient frontier determined by the solution of (6.23) will not be the same.

The hypothesis behind DRO is that the observed empirical distribution $\widehat{p}_T$ does not fully capture the full true statistical structure of the returns. Indeed, if one were able to to draw many samples from the true distribution of asset returns, each sample would capture a different view of the return structure. The optimization problem (6.22) would give a different solution for each sample. In reality, only a single historical realization is observed. The formulation (6.23) optimizes the expected value of the loss function under the worst–case probability distribution, chosen from all distributions whose distance from the observed empirical distribution is determined by the prespecified threshold ϵ.

14 Time independent in the sense that it is computed as an average over a window of T time steps. Of course, as the window advances in time, the values of the asset mean returns may change.

Equation (6.23) can be written

$$\min_{w,\gamma} \max_{q \in B_\epsilon(\widehat{p}_T)} \mathbb{E}_q \Big[\max \Big\{-\beta \bar{r}^\top w + (1-\beta)\gamma,$$
$$-\left(\beta + \frac{(1-\beta)}{\alpha}\right)\bar{r}^\top w + \left(1 - \frac{1}{\alpha}\right)(1-\beta)\gamma\Big\}\Big],$$
$$\text{subject to } e_n^\top w = 1,$$

which can be condensed as

$$\begin{aligned} &\min_{w,\gamma} \max_{q \in B_\epsilon(\widehat{p}_T)} \mathbb{E}_q \left[\max_{k=1,2} \left(a_k \bar{r}^\top w + b_k \gamma\right)\right], \\ &\text{subject to } e_n^\top w = 1, \\ &\text{with } a_1 = -\beta, \quad b_1 = 1 + a_1, \quad a_2 = -\beta - \frac{(1-\beta)}{\alpha}, \quad b_2 = 1 + a_2. \end{aligned} \tag{6.24}$$

Esfahani and Kuhn (2018, Corollary 5.1 and Section 7.1) have shown that, for the case in which the support of the true (unknown) probability distribution is a closed polytope, $\text{supp}(p(r)) = \{r \in \mathbb{R}^n | Cr \leq d\}$ for some $C \in \mathbb{R}^{m\times n}$ and $d \in \mathbb{R}^m$, then the DRO problem (6.24) can be approximated by the linear problem,

$$\min_{w,\gamma,\lambda,s_t,\gamma_{tk}} \lambda\epsilon + \frac{1}{T}\sum_{t=1}^{T} s_t,$$

such that

$$\begin{aligned} &e_n^\top w = 1, \\ &a_k\, r(t)^\top w + b_k \gamma + \gamma_{t,k}^\top (d - Cr(t)) \leq s_t, \quad && t = 1,\dots,T,\ k = 1,2, \\ &\|C^\top \gamma_{t,k} - a_k w\|_* \leq \lambda, && t = 1,\dots,T,\ k = 1,2 \\ &\gamma_{t,k} \geq 0, && t = 1,\dots,T,\ k = 1,2. \end{aligned} \tag{6.25}$$

In (6.25):

- $r(t)$, $t = 1,\dots,T$, are the observed historical returns;
- given a norm $\|v\|$ on $\mathbb{R}^n$, then $\|v\|_* = \sup_{\|r\|\leq 1} v^\top r$ is the dual norm of v;
- the Lagrange multiplier $\lambda \in \mathbb{R}$ enforces the constraint on the radius of the uncertainty set;
- the Lagrange multiplier vectors $\gamma_{t,k} \in \mathbb{R}^m$, $t = 1,\dots,T$, $k = 1,2$, enforce the constraint on the support of the probability distribution; and
- for a sample of values $s_t \in \mathbb{R}$, $t = 1,\dots,T$, $T^{-1}\sum_{t=1}^{T} s_t$ is the sample average.

Note that $\text{supp}(p(r))$ must include the support of the observed distribution of historical returns. Assuming that the constraint on the size of $\text{supp}(p(r))$ is observed regardless of whether it is imposed, we can assume $\gamma_{t,k} = 0$, $t = 1,\dots,T$, $k = 1,2$. We further use the $p = 1$ norm to define distance in $\mathbb{R}^n$ (and hence the KD $\mathcal{T}_1(\mu,\nu)$ defines the uncertainty set $B_\epsilon(\widehat{p}_T)$[15]), then $||\mathbf{v}||_{1,*} = ||\mathbf{v}||_\infty = \max_{i=1,\ldots,n} |v_i|$. Under this assumption and norm, the linear problem (6.25) reduces to

$$\begin{aligned}
&\min_{\mathbf{w},\gamma,\lambda,s_t} \ \lambda\epsilon + \frac{1}{T}\sum_{t=1}^{T} s_t,\\
&\text{such that}\\
&\quad \mathbf{e}_n^\top \mathbf{w} = 1,\\
&\quad a_k\,\mathbf{r}(t)^\top \mathbf{w} + b_k\gamma \le s_t, \qquad t = 1, \ldots, T,\ k = 1, 2,\\
&\quad \max_{i=1,\ldots,n} |-a_k w_i| = (-a_k)||\mathbf{w}||_\infty \le \lambda, \quad k = 1, 2.
\end{aligned} \tag{6.26}$$

Esfahani and Kuhn (2018) showed that the optimal solution for the vector of weights $\mathbf{w}$ converges to the equally weighted portfolio as ϵ tends to infinity. In section 6.3.2 we show numerical convergence to the equally weighted solution as ϵ increases.

The $\epsilon = 0$ limit of the problem (6.24) corresponds to the historical solution method of Chapter 4. However, this limit is not "handled well" by the linear problem (6.26). Setting $\epsilon = 0$ in (6.26) sets the term $\lambda\epsilon = 0$; the minimization over the multiplier λ and the last constraint are no longer needed, and (6.26) reduces to the classical problem. However, as long as $\epsilon > 0$ – no matter how small – the full system (6.26) is required. Thus, $\epsilon = 0$ is a degenerate limit of the system (6.26). Simply setting $\epsilon = 0$ in an existing code to solve (6.26) only sets $\lambda\epsilon = 0$; the minimization on λ and the final constraint remain. In our computations, this results in a solution with greater diversification of assets than that determined by historical optimization.

Long–only and long–short strategies can be implemented by the addition to (6.26) of the bound constraints

$$w_{\text{lb},i} \le w_i \le w_{\text{ub},i}, \qquad i = 1, \ldots, n. \tag{6.27}$$

6.3.1 Empirical Application of Robust CVaR Optimization using MATLAB

We present a MATLAB function `DOR_Mean_CVaR` that performs the optimization (6.26). This function was called from a `for` loop of a code that performed the moving window computation, updating hourly values of the asset weights. The first code listing provides a commented description of the input and output of the function as well as the indexing scheme for the vector of optimizing variables.

15 Note, Esfahani and Kuhn (2018) refer to $B_\epsilon(\widehat{p}_T)$ as the ambiguity set and $\text{supp}(p(\mathbf{r}))$ as the uncertainty set.

Listing 6.1: DRO_Mean_CVaR: Syntax

```
%{
Sol = DRO_Mean_CVaR(Rtn,alpha,beta,epsilon,cost_p,cost_m,x0)

Input Variables
Rtn: TxN matrix (doubles) of historical returns
       T is the number of observations
       N is the number of securities
alpha: scalar (double) is the confidence level of CvaR
beta: scalar (double) is the risk aversion coefficient
epsilon: scalar (double) is the radius of the uncertainty set
cost_p: scalar (double) is the transaction cost rate for asset
       purchase
cost_m: As per cost_p, but for asset sale
x0: column N-vector (doubles) of initial asset positions,
     in monetary units

Optimization Variables
The optimization variables are stored in a 2 + N + T column vector x
(doubles) having the components
x(1) = gamma
x(2) = lambda
x(3:N+2) = (w1, ..., wN)
x(N+3:N+T+2) = (s1, ..., sT)

Return value:
A structure Sol containing:
x_adj: column N-vector (doubles) of the new asset positions,
       in monetary units
weight: column N-vector (doubles) of the new asset weights
TO: scalar (double) the turnover of the portfolio
tot_cost: scalar (double) the total transactions cost
r_star: T-vector (doubles) portfolio expected returns for each
        historical date
mu_r: scalar (double) portfolio mean return
sig_r: scalar (double) portfolio standard deviation
fval: scalar (double) the minimized value of the objective function
exitflag: (integer) the exit flag from the linear programming solver
          linprog()
%}
```

The next code block comprises the start of the function, which defines fundamental variables and constants.

Listing 6.2: Global Variables and Constants

```
function Sol = DRO_Mean_CVaR(Rtn, alpha, beta, epsilon, cost_p, ...
                              cost_m, x0)

    % Set coefficients for DRO formulation
    a_k = [-beta; -beta - (1-beta)/alpha];
    b_k = [1-beta; (1-beta)*(1 - 1/alpha)];

    % Ensure x0 is a column vector, Rtn is a double array
    x0 = x0(:); % Convert to column vector
    Rtn = double(Rtn);

    % Extract Dimensions
    [T, N] = size(Rtn); % Correspond to 'T' and 'n' in (7.26)
    K = length(a_k); % Corresponds to 'k' in (7.26)

    % Values used repeatedly
    TK = T*K;
    NK = N*K;

    % Set number of minimization variables
    minvar = 2 + N + T;

    % Set number of constraint equations by type
    num_eq = 1; % sum_i(w_i) = 1
    num_ineq = TK + 2*NK ;
```

The linear programming problem (6.26) is to be solved using the MATLAB `linprog` function, which solves the generic linear problem

$$\min_{\boldsymbol{x}} \boldsymbol{f}^\top \boldsymbol{x}, \qquad \text{subject to} \begin{cases} \boldsymbol{A}_{\text{ineq}}\boldsymbol{x} & \leq \boldsymbol{b}_{\text{ineq}}, \\ \boldsymbol{A}_{\text{eq}}\boldsymbol{x} & = \boldsymbol{b}_{\text{eq}}, \\ \boldsymbol{x}_{\text{lb}} & \leq \boldsymbol{x} \leq \boldsymbol{x}_{\text{ub}}. \end{cases} \tag{6.28}$$

The relationship between (6.28) and (6.26) is easily established. As noted in Listing 6.1, the vector $\boldsymbol{x} = [\gamma, \lambda, w_1, \ldots, w_n, s_1, \ldots, s_T]^\top$. Consequently, the vector $\boldsymbol{f} = [0, \epsilon, 0, \ldots, 0, T^{-1}, \ldots, T$ The matrices $\boldsymbol{A}_{\text{eq}}$ and $\boldsymbol{A}_{\text{ineq}}$ must be written using column indexing defined by the vector $\boldsymbol{x}$. To make the final inequality constraint in (6.26) linear, it can be rewritten as the $2n$ constraints,

$$|-a_k w_i| \leq \lambda, \qquad i = 1, \ldots, n, \; k = 1, 2,$$

or equivalently,

$$-\lambda \le -a_k w_i \le \lambda, \qquad i = 1, \ldots, n, \ k = 1, 2,$$

which can further be expressed as the $4n$ constraints

$$a_k w_i - \lambda \le 0, \quad -a_k w_i - \lambda \le 0, \qquad i = 1, \ldots, n, \ k = 1, 2. \tag{6.29}$$

Using the notation established by Listing 6.2, $\boldsymbol{A}_{\text{eq}}$ is a $1 \times$ `minvar` matrix. Employing the constraint form (6.29), $\boldsymbol{A}_{\text{ineq}}$ is a $(TK + 2NK) \times$ `minvar` matrix. The vectors $\boldsymbol{b}_{\text{eq}}$ and $\boldsymbol{b}_{\text{ineq}}$ are, respectively, 1×1 and $(TK + 2NK) \times 1$.

Listing 6.3 computes $\boldsymbol{f}$, $\boldsymbol{A}_{\text{ineq}}$, $\boldsymbol{b}_{\text{ineq}}$, $\boldsymbol{A}_{\text{eq}}$, $\boldsymbol{b}_{\text{eq}}$, $\boldsymbol{x}_{\text{lb}}$, and $\boldsymbol{x}_{\text{ub}}$. Only the w_i components of $\boldsymbol{x}_{\text{lb}}$ and $\boldsymbol{x}_{\text{ub}}$ are bounded; no bound is placed on its γ, λ, and s_t, $t = 1, \ldots, T$ components. The listing organizes these arrays in a structure

```
prob = struct(
    f, ...
    Aineq, ...
    bineq, ...
    Aeq, ...
    beq, ...
    lb, ...
    ub);
}
```

While this structure is not necessary given the current form of the function, it is consistent with an object–oriented programming approach. It would also provide a convenient way of passing the problem details to the calling routine via the addition of an entry `Sol.prob` in the return structure.

The matrices $\boldsymbol{A}_{\text{ineq}}$ and $\boldsymbol{A}_{\text{eq}}$ are stored in sparse matrix format, which lists non–zero entries in the form i, j, A_{ij} The formulation of these i, j indexes employs a faily sophisticated used of the MATLAB functions `repmat`, `repelem`, `kron`, and `reshape`. The reader can consult the relevant online MATLAB Help Center pages for detail on these functions.

Listing 6.3: Problem Formulation

```
% Create a structure to store the problem description
prob = struct();

%%%% Build the constraint matrices

% Inequality constraint A
% b_k*gam + a_k*<w,r_t> - s_t <= 0

% Block 1: b_k*gam terms (T*K elements)
```

```
subi_1 = transpose(1:TK); % column vector of indices 1, ..., TK
subj_1 = ones(TK, 1); % gam is variable 1
 cof_1 = [repmat(b_k(1), T, 1); repmat(b_k(2), T, 1)];

% Block 2: a_k*<w,r_t> terms (T*K*N elements)
subi_2 = repelem( subi_1, N );
subj_2 = repmat( transpose(3:N+2), TK, 1); % w variables
 cof_2 = reshape( transpose(kron(a_k, Rtn)), T*N*K, 1, 1);

% Block 3: -s_t terms (T*K elements)
subi_3 = subi_1;
subj_3 = N + 2 + tranpose( repmat(1:T, 1, K) );
 cof_3 = -subj_1;

% Form constraint A
subi_A = [ subi_1; subi_2; subi_3 ];
subj_A = [ subj_1; subj_2; subj_3 ];
 cof_A = [ cof_1; cof_2; cof_3 ];

clear subi_1 subj_1 cof_1
clear subi_2 subj_2 cof_2
clear subi_3 subj_3 cof_3

% Inequality constraint B
% a_k*w - lambda <= 0 and -a_k*w - lambda <= 0

% Block 1: -lambda terms (2*N*K elements)
   tmp = ones(2*NK,1);
subi_1 = transpose(TK+1:TK + 2*NK);
subj_1 = 2*tmp; % lambda is variable 2
 cof_1 = -tmp;

% Block 2: +/- a_k*w terms (2*NK elements)
subi_2 = subi_1;
subj_2 = repmat( transpose(3:N+2), 2*K, 1 );
 cof_2 = repelem(a_k, N);
 cof_2 = [cof_2; -cof_2];

% Form constraint B
subi_B = [ subi_1; subi_2 ];
subj_B = [ subj_1; subj_2 ];
 cof_B = [ cof_1; cof_2 ];
```

```
clear subi_1 subj_1 cof_1 tmp
clear subi_2 subj_2 cof_2
clear subi_3 subj_3 cof_3

 % Combine constraints A and B
subi = [ subi_A; subi_B ];
subj = [ subj_A; subj_B ];
 cof = [ cof_A; cof_B ];

clear subi_A subj_A cof_A
clear subi_B subj_B cof_B

% Store Aineq in sparse matrix form
prob.Aineq = sparse(subi, subj, cof, num_ineq, minvar);

% Equality constraint C
% sum_i(w_i) = 1
subi_C = ones(N,1);
subj_C = transpose(3:N+2);
 cof_C = subi_C;

prob.Aeq = sparse(subi_C, subj_C, cof_C, num_eq, minvar);

clear subi_C subj_C cof_C

% Set constraint right-hand-side vectors
prob.bineq = [ +zeros(TK + NK, 1); +zeros(NK, 1) ];
prob.bineq = [ +zeros(TK + NK, 1); +zeros(NK, 1) ];

% Set bounds on optimization variables
 prob.lb = -inf(minvar, 1);
 prob.ub = inf(minvar, 1);
 prob.lb(3:N+2) = 0; % defines long-only
 prob.ub(3:N+2) = 1; % maximum weight = 1

% Set objective function
prob.f = [0; epsilon; zeros(N, 1); (1/T)*ones(T, 1)];
```

The next listing, 6.4, attempts a solution of the linear programming problem via `linprog` using the `dual-simplex` solver. If the solver fails to find a solution, a warning message is printed and the `exitflag` and final objective function value `fval` are passed, via the structure Sol, to the calling routine for a decision on how to proceed. Solver failure might be handled locally in this function using the try { ...} catch { ...} end MATLAB functionality. As the form the try and catch statements can be problem dependent, we do not consider this approach in the current listing.

Listing 6.4: Problem Solution

```
% Set linprog() ptions
options = optimoptions('linprog', 'Display', 'iter', 'Algorithm', ...
                    'dual-simplex');

% Solve with linprog ()
[x,fval,exitflag] = linprog(prob.f,prob.Aineq,prob.bineq, ...
             prob.Aeq,prob.beq,prob.lb,prob.ub,options);

% Check result
if exitflag ~= 1
    warning('\nOptimization failed to converge; exitflag %d\n',...
             exitflag);
end
```

The final listing, 6.5, computes the variables to be returned by the function.

Listing 6.5: Computing Output Values

```
% Extract the optimal weights from the optimized solution x
weight = x(3:N+2);

% Calculate asset position values in monetary units
x0_tot = sum(x0);
x = weight * x0_tot;

% Compute buy and sell transaction costs
delta = x - x0;
c_p = delta( delta > 0 ) * cost_p;
c_m = delta( delta < 0 ) * cost_m;

% Total cost
tot_cost = sum(c_p) + abs( sum(c_m) );

% The portfolio must fund its own maintenance costs
% Adjust the portfolio positions
x_adj = weight * ( x0_tot - tot_cost );

% Compute the turnover on the adjusted value
```

```
    % (this includes transaction costs)
    TO = 0.5*sum( abs(x_adj - x0) ) / x0_tot;

    % Compute portfolio expected value and standard deviaion
    r_star = Rtn * weight;
      mu_r = mean(r_star);
     sig_r = std(r_star);

    % Output structure
    Sol = struct(...
        'Portfolio', x_adj, ...
        'Weights' , weight, ...
        'Turnover' , TO, ...
        'Transcost', tot_cost, ...
        'r_star' , r_star, ...
        'mu_r' , mu_r, ...
        'sigma_r' , sig_r, ...
        'opt_sol' , fval, ...
        'exit' , exitflag);
end
```

6.3.2 Empirical Examples: Robust CVaR Optimization

We present numerical results for the DRO formulation of the mean–CVaR problem (6.26).[16] We test the method under the same setting as in Chapter 4, using a rolling window of 2000 hourly return data points to fit the empirical distribution function. Portfolio strategies are evaluated through performance measures over the out-of-sample period, from 22:00:00 on October 20, 2021 to 13:00:00 on December 1, 2021. We consider a long–only strategy, adding the set of constraints $w_i \geq 0$, $i = 1, \ldots, n$, to (6.26).

We concentrate on the effects of changing the radius ϵ of the Kantorovich ball and the risk–aversion parameter β. To illustrate these effects with greater clarity, we do not impose turnover constraints. To test the effect of changing the radius ϵ, we apply the Herfindahl–Hirschman index to the time–varying weight solutions $\boldsymbol{w}(t) = \{w_1(t), \ldots, w_n(t)\}$,

$$\mathrm{HHI}_t = \sum_{i=1}^{n} w_i^2(t). \tag{6.30}$$

16 Using the $p = 1$ norm, we tested solutions of (6.26) against those of the larger problem (6.25). Both yield the same solutions, reinforcing the use of our assumption $\gamma_{t,k} = 0$, $t = 1, \ldots, T$, $k = 1, 2$.

We will work with a time–averaged index

$$\mathrm{HHI}_{t_1,t_2} = \frac{1}{(t_2 - t_1)} \sum_{t=t_1}^{t_2} \mathrm{HHI}_t. \tag{6.31}$$

When the time interval $[t_1, t_2]$ is known from context, we refer to the time–averaged index simply as HHI. If the portfolio is perfectly diversified over $[t_1, t_2]$ (all n assets have the same weight, as in the EWP), HHI will be equal to $n/n^2 = 1/n$. On the other hand, if the portfolio is highly concentrated on a few assets, the HHI will be larger, approaching 1 in the extreme case where a single asset holds all the weight for the entire time interval.

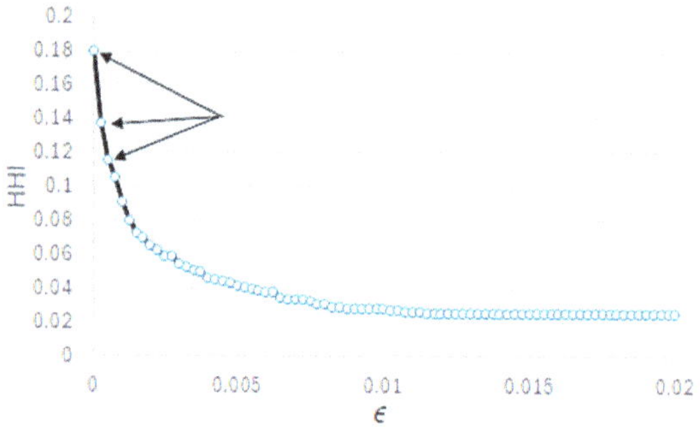

Fig. 6.3: HHI as a function of ϵ when $\beta = 0$. Arrows indicate the values for $\epsilon = \{0, 2.5 \cdot 10^{-4}, 5 \cdot 10^{-4}\}$

Recall that $\epsilon = 0$ corresponds to the historical solution method of Chapter 4, $\beta = 0$ corresponds to minimizing CVaR_α, while the choice $\beta = 1$ corresponds to a greedy investor who only seeks to maximize return. To determine a range of values of ϵ to investigate, we considered the DRO CVaR_{95} optimization under a long–only strategy with $\beta = 0$ and a range of choices of ϵ. For each value of ϵ, HHI was computed as an average over the 1000 hourly optimizations from 22:00:00 on October 20, 2021 to 13:00:00 on December 1, 2021. Fig. 6.3 presents the observed dependence of HHI on ϵ. By $\epsilon = 0.1$, HHI was within 0.1% of the equi–weighted value of 0.025. For our numerical studies, we therefore chose to use the three values $\epsilon = \{0, 2.5 \cdot 10^{-4}, 5 \cdot 10^{-4}\}$, which correspond to the range where HHI is changing most rapidly. For $\beta = 0$, this choice corresponds to the HHI values indicated by the three arrows in Fig. 6.3.

Fig. 6.4 presents the cumulative return for the DRO CVaR_{95} and CVaR_{99} optimizations under the long–only strategy for several choices of the parameter pair (β, ϵ). Also shown is the performance of the EWP benchmark. For $\beta = 0.2$ and 0.3, under CVaR_{95} optimization the larger value of $\epsilon = 2.5 \cdot 10^{-4}$ produced lower returns compared to $\epsilon = 0$. For the two values of $\beta = 0.3$ and 0.5 shown for the CVaR_{99} optimization, the larger value of $\epsilon = 2.5 \cdot 10^{-4}$ produced higher returns than $\epsilon = 0$.

To test whether the DRO approach, i.e., $\epsilon > 0$, does improve performance, detailed statistics for the case $\alpha = 0.95$ and choices of (β, ϵ) are reported in Tab. 6.1. For the 1,000

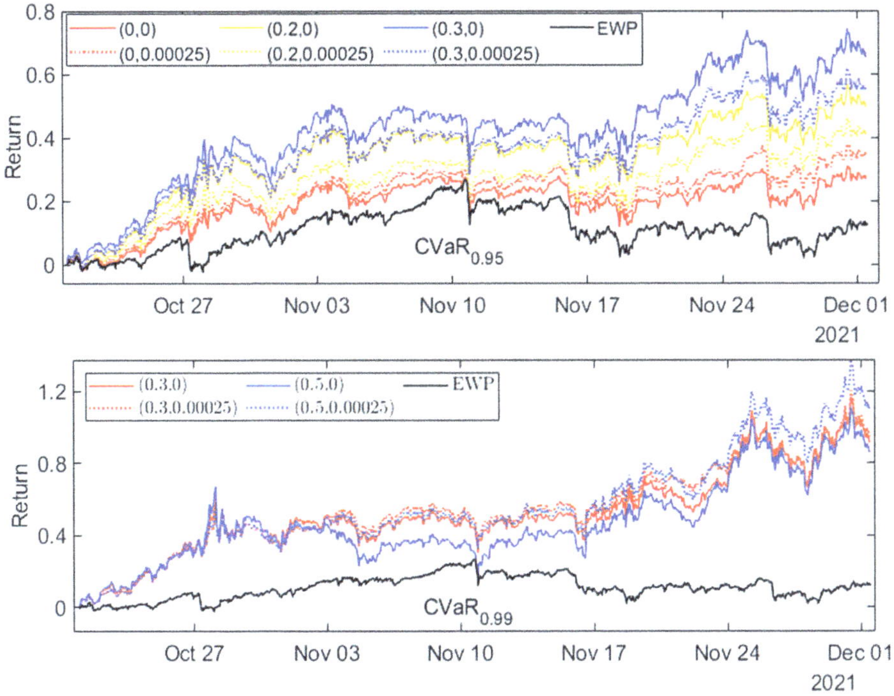

Fig. 6.4: Cumulative return of DRO $CVaR_{95}$ and $CVaR_{99}$ optimizations under the long-only strategy with different combinations of (β, ϵ). The return for the EWP benchmark is included for comparison

hour out-of-sample period, the table reports the annualized return (AnnRet), the total turnover (TO), the average daily HHI, and the total SS and RR ratios. We first consider the dependence on ϵ under constant values of β and look for clear trends in the tabulated data. With only three values of ϵ per β, the ability to see trends is highly reduced. For all four values of β, there is a reduction in HHI as ϵ increases, indicating a trend toward diversification. For AnnRet, SS and RR there is a trend of reduction in value as ϵ increases, but only for $\beta = 0.2$ and 0.3. For the smaller values of β there are no obvious trends. There are no consistent trends of TO with ϵ.

If we consider dependence on β under constant values of ϵ, there appears to be a clear trend that AnnRet, HHI, SS and RR increase with β for $\beta \geq 0.1$. However, for $\beta = 0$ the trend is not always followed. This may be explained by the fact that higher values of β indicate a more risk-seeking attitude; the portfolio manager places more emphasis on the expected value of the portfolios. There is the suggestion that TO increases with $\beta \in \{0, 0.1, 0.2\}$, with TO then decreasing for $\beta = 0.3$.
Finally, we note that the optimized portfolios significantly outperformed the benchmark EWP and EWBH strategies in terms of AnnRet, HHI, SS and RR. However, the turnover of the optimized portfolios is over 10 times worse than that of the EWP.

Tab. 6.1: Summary statistics for out-of-sample hourly DRO $CVaR_{95}$ optimizations under the long-only strategy from 22:00:00 on October 20, 2021 to 13:00:00 on December 1, 2021.

Optimization		AnnRet (%)	TO ($\times 10^1$)	HHI ($\times 10^{-1}$)	SS ($\times 10^{-2}$)	RR
	EWBH	5.79	0.00	0.332	4.23	0.869
	EWP	3.09	0.14	0.250	2.57	0.861
β	ϵ ($\times 10^{-4}$)					
0.0	0.0	6.21	2.12	1.65	4.77	0.971
0.0	2.5	7.70	2.04	1.12	5.52	0.953
0.0	5.0	7.94	2.13	1.04	5.73	0.956
0.1	0.0	7.78	2.29	1.54	5.53	0.964
0.1	2.5	7.89	2.08	1.18	5.57	0.956
0.1	5.0	7.82	2.12	1.10	5.58	0.960
0.2	0.0	10.8	2.13	1.59	6.90	0.982
0.2	2.5	9.08	2.35	1.34	6.06	0.969
0.2	5.0	8.49	2.61	1.16	5.91	0.965
0.3	0.0	13.7	1.93	1.77	7.96	1.006
0.3	2.5	11.7	2.29	1.51	7.24	0.980
0.3	5.0	10.4	2.27	1.33	6.73	0.968

Tab. 6.2 shows the detailed statistics for the DRO $CVaR_{99}$ optimization. The table has an extra column to quantify transaction costs. Specifically, the column displays the total relative cost (TRC) of the portfolio. The TRC is defined as the sum of all transaction costs over the entire out-of-sample period expressed as a fraction of the initial investment. Transaction costs were computed assuming a rate of 2 basis points on each (separate buy and sell) transaction.

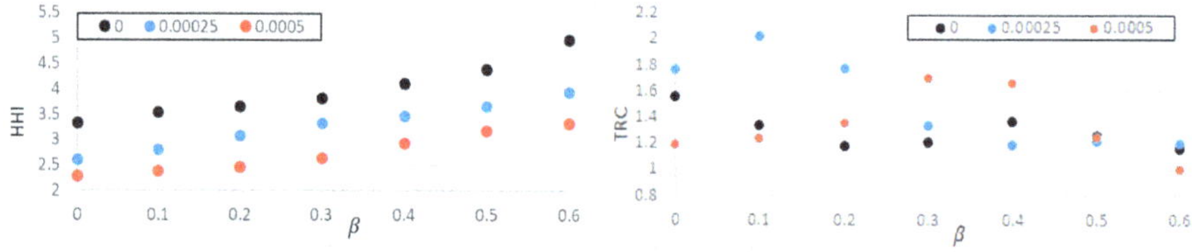

Fig. 6.5: HHI and TRC as a function of β at constant values of ϵ from the data in Tab. 6.2

Except for HHI, there are no discernible trends in each statistic as β increases under constant value of ϵ. Fig. 6.5 shows the trend in HHI. There is a monotonic increase in HHI (portfolios become more concentrated on fewer assets) with β at constant ϵ. The plot also shows that HHI decreases (portfolios become more diversified) as ϵ increases

Tab. 6.2: Summary statistics for out-of-sample hourly DRO $CVaR_{99}$ optimizations under the long-only strategy from 22:00:00 on October 20, 2021 to 13:00:00 on December 1, 2021.

Optimization		AnnRet (%)	TRC (%)	TO ($\times 10^{1}$)	HHI ($\times 10^{-1}$)	SS ($\times 10^{-2}$)	RR
	EWBH	5.79	0.00	0.00	0.332	4.23	0.869
	EWP	3.09	0.12	0.14	0.250	2.55	0.861
β	ϵ ($\times 10^{-4}$)						
0.0	0.0	15.67	1.55	2.57	3.338	7.53	1.049
0.0	2.5	16.97	1.76	3.03	2.611	8.76	1.054
0.0	5.0	15.72	1.18	2.13	2.293	8.47	1.041
0.1	0.0	16.97	1.33	2.20	3.538	7.93	1.071
0.1	2.5	17.15	2.02	3.36	2.815	8.71	1.069
0.1	5.0	16.20	1.23	2.14	2.388	8.63	1.049
0.2	0.0	15.88	1.17	2.04	3.656	7.38	1.066
0.2	2.5	17.22	1.77	2.99	3.095	8.46	1.057
0.2	5.0	18.01	1.35	2.32	2.478	9.43	1.080
0.3	0.0	17.78	1.20	2.07	3.825	7.94	1.076
0.3	2.5	18.27	1.33	2.19	3.339	8.74	1.070
0.3	5.0	16.45	1.70	2.92	2.653	8.59	1.087
0.4	0.0	18.76	1.36	2.34	4.108	8.15	1.110
0.4	2.5	19.63	1.18	1.96	3.483	9.10	1.078
0.4	5.0	17.27	1.66	2.78	2.937	8.66	1.076
0.5	0.0	16.88	1.25	2.29	4.391	7.17	1.094
0.5	2.5	20.45	1.21	2.05	3.671	9.20	1.079
0.5	5.0	17.84	1.24	2.04	3.190	8.77	1.089
0.6	0.0	14.78	1.15	2.16	4.970	5.73	1.072
0.6	2.5	20.94	1.19	2.06	3.958	9.15	1.093
0.6	5.0	18.77	1.00	1.67	3.332	9.08	1.105

at constant β. For comparison Fig. 6.5 also displays the dependence of TRC on β. While the $\beta = 0.5$ and 0.6 values hint of a decrease of TRC at large values of beta, there are no discernible trends in the data.

The increase in portfolio diversification as ϵ increases is demonstrated in Fig. 6.6 which shows the time dependent weight composition of the DRO $CVaR_{99}$ optimized portfolio under a long-only strategy, comparing the results for $\epsilon = 0$ and $\epsilon = 5 \times 10^{-4}$ when $\beta = 0.3$. Also plotted is the average weight of each asset over the entire time period. At the higher value of ϵ, the assets ALGO, AXS, EGLD and NEAR are added to the portfolio in greater quantities. It is interesting to compare the weight compositions in Fig. 6.6 with those of the TVP and T99 historical optimizations in Fig. 4.2.

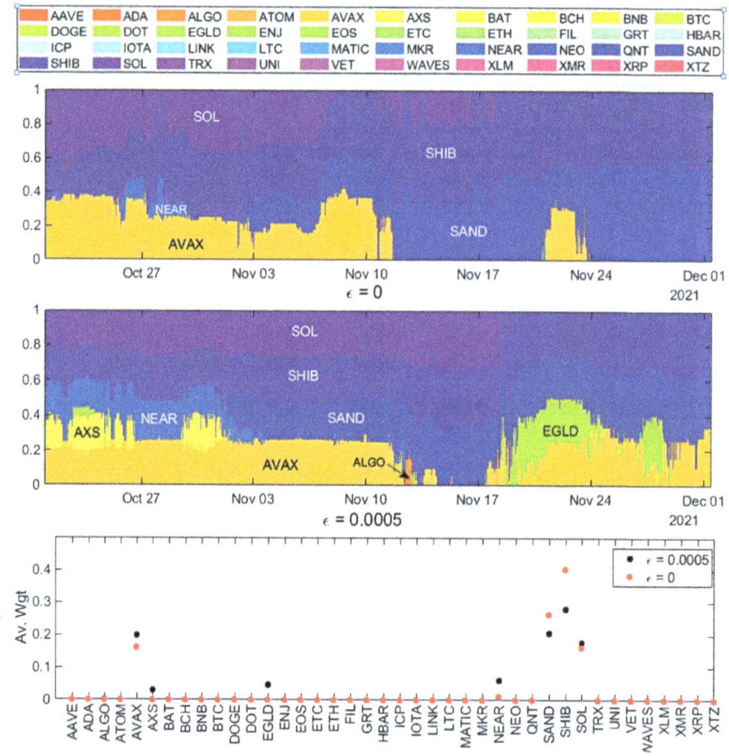

Fig. 6.6: Time dependent weight composition of the DRO CVaR_{99} optimization under a long-only strategy. The optimization parameters were $\beta = 0.3$ and $\epsilon = \{0, 5 \cdot 10^{-4}\}$. Also shown is the average weight of each asset over the entire time period.

7 Backtesting

Starting in 1998, US regulatory guidelines required that banks with significant trading activity set aside capital (the market–risk capital requirement) to guard against extreme portfolio losses (Campbell 2005). The size of this capital requirement is determined by the value–at–risk (VaR_α) of the portfolio at a pre–specified confidence level of $1 - \alpha$. VaR_α is defined in (3.23). As noted regarding CVaR notation in the last paragraph of Section 3.2.3, it is common to write $\mathrm{VaR}_{100(1-\alpha)}$, using the confidence level $100(1 - \alpha)$, rather than VaR_α. We will employ that notation here, writing VaR_{99} for $\mathrm{VaR}_{0.01}$, etc.

Let $\mathrm{VaR}_{99}(t + 10)$ denote an assessment of a bank's 1% VaR over the next 10 trading days (a 10–day horizon beyond day t). The market–risk capital (MRC) requirement of a bank is then set (Campbell 2005) as the greater of the current horizon estimate and a multiple of the average value of the bank's reported values of $\mathrm{VaR}_{99}(t + 10)$ over the previous 60 trading days, plus an additional amount:

$$\mathrm{MRC}(t) = \max\left(\mathrm{VaR}_{99}(t + 10),\ S(t) \cdot \frac{1}{60} \sum_{i=0}^{59} \mathrm{VaR}_{99}(t - i + 10) \right) + c.$$

The additional amount c reflects the underlying credit risk of the bank's portfolio, and is bank specific. The multiplying factor $S(t)$ is determined through backtesting by classifying the number x of violations of the VaR_{99} value over the preceding 250 trading days into three distinct categories. These categories, provided in Tab. 7.1, define the Basel II green, yellow, and red (traffic light) zones. The last column of the table provides the binomial cumulative probability distribution, $F_\mathrm{B}(X \le x \mid 250, 0.01)$, for observing no more than x violations of the VaR_{99} value in the previous 250 days. It represents the cumulative probability of seeing x or fewer violations due to chance alone.[1]

Risk–based capital requirements depend on both portfolio performance and the bank's VaR model, which are interdependent. To meet MRC requirements, accurate VaR estimation and portfolio performance testing via VaR are essential. Portfolio returns are treated as random variables, and at time t, the distribution governing these returns is characterized by a VaR at the desired confidence level (e.g., 99%). As noted by Christoffersen (1998), if portfolio management responds adequately to market–risk conditions and the VaR model is sufficiently accurate, the sequence of VaR violations should satisfy two key properties:

1. *Unconditional coverage:* The probability of a violation (exceeding $\mathrm{VaR}_\alpha(t)$) should equal α. This property restricts how often violations occur.
2. *Independence:* The probability of a violation at time t should be independent of when previous violations occurred. In other words, the history of violations should

1 Assuming that the occurrence of violations is governed by a binomial distribution with probability 0.01, and that observations of violations over time (that is, samples from the distribution) are iid.

https://doi.org/10.1515/9781501517136-007

Tab. 7.1: Basel II guidelines for $S(t)$

Zone	x	$S(t)$	$F_B(X \leq x \mid 250, 0.01)$
	0	3.00	0.0811
	1	3.00	0.2858
Green	2	3.00	0.5432
	3	3.00	0.7581
	4	3.00	0.8922
	5	3.40	0.9588
	6	3.50	0.9863
Yellow	7	3.65	0.9960
	8	3.75	0.9989
	9	3.85	0.9997
Red	≥ 10	4.00	0.9999

not influence future occurrences. Since market volatility often clusters, independence requires not only an accurate VaR model but also rapid, risk–adjusted portfolio management.

Selected VaR–based standard tests, detailed in Section 7.1, evaluate these properties. Under Basel II, testing based on VaR_{95} is required; while Basel III mandates testing based on VaR_{99}. Additionally, this chapter explores testing at the 99.5% confidence level ($\mathrm{VaR}_{99.5}$).

The bank capital requirements described at the beginning of this chapter require VaR computations based upon returns determined over a trading day. This chapter demonstrates VaR backtesting on historical optimizations of Chapter 4, dynamic optimizations of Chapter 5 and distributionally robust optimizations of Chapter 6. As out data sets employ hourly returns, we describe VaR backtesting based on hourly returns. The discussion in this chapter can be adapted to return data gathered over any other time interval (minute returns, 24–hour returns, monthly returns, etc.) by replacing "hour" with the appropriate time interval.

The core question in backtesting on our cryptocurrency data is whether the portfolio return $r_p(t)$ for the current hour t exceeds the projected $\mathrm{VaR}_\alpha(r_p, t)$ for the same hour. Such occurrences, termed "failures" or violations, are known by the end of the hour. The primary challenge in backtesting lies in estimating $\mathrm{VaR}_\alpha(r_p, t)$ for hour t, estimation of which typically relies on the portfolio's performance over a prior τ–hour period.

For historical and distributionally robust optimizations, $\mathrm{VaR}_\alpha(r_p, t)$ is computed empirically based on synthetic portfolio returns from hours $t - \tau, \dots, t - 1$. Let $r_{i,k}$, $k = t - \tau, \dots, t - 1$, $i = 1, \dots, n$, denote the individual asset returns during this period, and $w_i(t)$, $i = 1, \dots, n$, the optimized weights for hour t. The portfolio returns $r_{p,k} = \sum_{i=1}^{n} w_i(t) r_{i,k}$, $k = t - \tau, \dots, t - 1$, represent expected returns for hour t. The values of $\mathrm{VaR}_\alpha(r_p, t)$ are estimated based on the distribution of these synthetic returns.

We used $\tau = 2,000$, which aligns the VAR computation with the rolling windows used for the historical and distributionally robust optimizations.

For dynamic optimizations, a large sample of dynamic returns $\{r_{i,s,t},\ s = 1,\dots,S\}$, $S = 10,000$, was computed for each asset $i = 1,\dots,n$ in the portfolio for hour t. Applying the optimized weights $w_i(t)$, $i = 1,\dots,n$, to these returns yields a sample of predicted portfolio returns $r_{p,s,t} = \sum_{i=1}^{n} w_i(t) r_{i,s,t}$, $s = 1,\dots,S$. The values of $\mathrm{VaR}_\alpha(r_p,t)$ are then estimated based on the distribution of these predicted returns.

7.1 VaR Tests

Consider individual asset return data for the hours $1,\dots,T$. As our optimization methods utilize a moving window of τ hours to calculate optimized portfolio weights, the backtest dataset consists of $N = T - \tau$ values, corresponding to hours $\tau+1,\dots,T$. These values represent observed portfolio returns and estimated VaR, or equivalently, a time series of N outcomes, where each outcome is classified as either a "success" or a "failure." Standard backtesting procedures apply various tests to this dataset.

7.1.1 Binomial Test

For each of the N hours, either the (negative of the) portfolio return exceeds the $\mathrm{VaR}_\alpha(r_p,t)$ value (a failure) or it does not (a success). The number of failures is expected to follow a binomial distribution, with α representing the probability of failure. Therefore, the expected number of failures is $N\alpha$, with standard deviation of $\sqrt{N\alpha(1-\alpha)}$. Let x denote the observed number of failures. Using the z–score as the test statistic,

$$z(x) = \frac{x - N\alpha}{\sqrt{N\alpha(1-\alpha)}}, \tag{7.1}$$

and assuming N, $N\alpha$, and $N(1-\alpha)$ are sufficiently large,[2] the z–score approximately follows the standard normal distribution $\mathcal{N}(0,1)$.

The (tail) probability that this z–score is exceeded is given by $1 - F_{\mathcal{N}}(z(x)$, where $F_{\mathcal{N}}(\cdot)$ is the CDF of the standard normal distribution. As the binomial test (BIN) is a two–sided test, the p–value $p_{\mathrm{BIN}}(z(x))$ is twice the tail probability. The binomial test accepts the null hypothesis, H_0: the probability of failure is α, if

$$p_{\mathrm{BIN}}(z(x)) = 2\left[1 - F_{\mathcal{N}}(z(x))\right] > p_{\mathrm{test}}, \tag{7.2}$$

where $1 - p_{\mathrm{test}}$ is the test confidence level, typically set to 0.95; that is, $p_{\mathrm{test}} = 0.05$.

2 As long as $\alpha \neq 0$, we have $N\alpha = \mathcal{O}(N)$ and $N(1-\alpha) = \mathcal{O}(N)$ for sufficiently large N. In practice (based on the three standard deviation confidence interval for the point estimate x/N), it suffices to ensure $N \geq 9\max\left((1-\alpha)/\alpha,\ \alpha/(1-\alpha)\right)$.

7.1.2 Traffic Light Test

The Basel framework for backtesting introduces the traffic light (TL) test, which is also based on a binomial distribution. As described in Section 7.1.1, N represents the number of observations, α is the probability of failure, and x is the observed number of failures. The probability of observing up to x failures in N trials (samples) is

$$P(X \leq x \mid N, \alpha) = F_{\mathrm{B}}(X \leq x \mid N, \alpha). \tag{7.3}$$

where $F_{\mathrm{B}}(x \mid N, \alpha)$ denotes the binomial CDF with parameters N and α. The Basel framework classifies outcomes into the three zones given in Tab. 7.1. The table is based on a sample of $N = 250$ observations. For other sample sizes, the exact zone boundaries are

$$\begin{aligned} &\text{Green:} & & F_{\mathrm{B}}(X \leq x \mid N, \alpha) < 0.95, \\ &\text{Yellow:} & 0.95 \leq\ & F_{\mathrm{B}}(X \leq x \mid N, \alpha) < 0.9999, \\ &\text{Red:} & 0.9999 \leq\ & F_{\mathrm{B}}(X \leq x \mid N, \alpha). \end{aligned} \tag{7.4}$$

The tests defined in (7.4) are collectively referred to as the TL test.

The Basel framework introduces the (time dependent) multiplication factor S to determine the potential increase in a firm's capital requirements (relative to a baseline) based on the TL test results. This factor can be re-expressed as a (time dependent) scaling factor $s = S - 3$, which equals 0 for the green zone, 1 for the red zone, and increases within the yellow zone depending on the difference between the assumed VaR quantile level $1 - \alpha$ and the observed quantile level $1 - x/N$. Under the assumption of a standard normal distribution, the z–score (z_{norm}) corresponding to the cumulative value $1 - \alpha$ and the z–score(z_{obs}) corresponding to the cumulative value $1 - x/N$ can be calculated. The scaling factor s_{yellow} for the yellow zone is then determined as

$$s_{\mathrm{yellow}} = 3\left(\frac{z_{\mathrm{norm}} - z_{\mathrm{obs}}}{z_{\mathrm{obs}}}\right), \quad \text{s.t. } 0 \leq s_{\mathrm{yellow}} \leq 1. \tag{7.5}$$

7.1.3 Kupiec's Tests

Kupiec (1995) introduced two tests: the proportion of failures (PoF) test and the time until first failure (TUFF) test.

The PoF test evaluates whether the point prediction x/N aligns with the expected failure probability α. Assuming the number of failures in N trials follows a binomial distribution with failure probability α, the probability of observing x failures is given by

$$P(x \mid N, \alpha) = \binom{N}{x} \alpha^x (1 - \alpha)^{N-x}.$$

The PoF test determines whether x/N is consistent with α by analyzing the likelihood ratio

$$R_{\mathrm{PoF}} = \frac{\binom{N}{x}\alpha^x(1-\alpha)^{N-x}}{\binom{N}{x}(\frac{x}{N})^x(1-\frac{x}{N})^{N-x}} = \frac{\alpha^x(1-\alpha)^{N-x}}{(\frac{x}{N})^x(1-\frac{x}{N})^{N-x}}.$$

Equivalently, the test examines the logarithm of this ratio (the log–likelihood ratio)

$$LR_{\text{PoF}} = -2\ln R_{\text{PoF}} = -2\left[x\ln\left(\frac{N\alpha}{x}\right) + (N-x)\ln\left(\frac{N(1-\alpha)}{N-x}\right)\right].$$

This statistic has the limiting values,

$$\lim_{x\to 0} LR_{\text{PoF}} = -2N\ln(1-\alpha),$$
$$\lim_{x\to N} LR_{\text{PoF}} = -2N\ln\alpha.$$

For large N, LR_{PoF} follows a χ^2–distribution with one degree of freedom. The p–value of the PoF test, denoted p_{PoF}, is

$$p_{\text{PoF}} = 1 - F_{\chi_1^2}(LR_{\text{PoF}}), \tag{7.6}$$

where $F_{\chi_1^2}(\cdot)$ is the CDF of the χ^2–distribution with one degree of freedom. The PoF test accepts the null hypothesis, H_0: x/N is consistent with α, if

$$p_{\text{PoF}} > p_{\text{test}}, \tag{7.7}$$

where p_{test} is a user–defined significance level, typically 0.05 (corresponding to a 95% confidence level).

The TUFF test evaluates whether the number of hours, n_1, until the first failure is consistent with α. Similar to the PoF test, it employs a log–likelihood ratio statistic:

$$LR_{\text{TUFF}} = -2\ln\left(\frac{\alpha(1-\alpha)^{n_1-1}}{(\frac{1}{n_1})(1-\frac{1}{n_1})^{n_1-1}}\right) = -2\left[\ln(n_1\alpha) + (n_1-1)\ln\left(\frac{n_1(1-\alpha)}{n_1-1}\right)\right]. \tag{7.8}$$

This statistic has the limiting value

$$\lim_{n_1\to 1} LR_{\text{TUFF}} = -2\ln\alpha.$$

For large N, LR_{TUFF} also follows a χ^2–distribution with one degree of freedom. The p–value for the TUFF test is

$$p_{\text{TUFF}} = 1 - F_{\chi_1^2}(LR_{\text{TUFF}}). \tag{7.9}$$

The TUFF test accepts the null hypothesis, H_0: n_1 is consistent with α, if

$$p_{\text{TUFF}} > p_{\text{test}}. \tag{7.10}$$

If no failures are observed ($n_1 = 0$), LR_{TUFF} is undefined. In this case, two scenarios are considered.

1. If $N > \alpha^{-1}$ and if the TUFF test fails when $n_1 = N + 1$ (the earliest possible value of n_1 given no failures in N hours), the the null hypothesis is rejected. Values of LR_{TUFF} and p_{TUFF} are the reported for $n_1 = N + 1$.
2. Otherwise, it is not possible for TUFF test to accept or reject the null hypothesis.

7.1.4 Christoffersen's Tests

Christoffersen (1998) proposed the conditional coverage independence (CCI) test, which evaluates whether the probability of observing a failure at hour t depends on whether a failure occurred at the previous hour $t-1$. This test examines all possible pairs of hours, $(t-1, t)$, and computes:

n_{00}: the number of pairs where no failure occurred at both hours;

n_{10}: the number of pairs where a failure occurred at hour $t-1$, followed by no failure at hour t;

n_{01}: the number of pairs where no failure occurred at hour $t-1$, followed by a failure at hour t; and

n_{11}: the number of pairs where a failure occurred at both hours.

From these counts, the following probabilities are estimated:

$\pi_0 = n_{01}/(n_{00}+n_{01})$: the conditional probability of a failure at hour t given no failure at hour $t-1$;

$\pi_1 = n_{11}/(n_{10}+n_{11})$: the conditional probability of a failure at hour t given a failure at hour $t-1$; and

$\pi = (n_{01}+n_{11})/(n_{00}+n_{01}+n_{10}+n_{11})$: the unconditional probability of a failure at hour t.

The CCI test uses the log–likelihood ratio test statistic

$$\begin{aligned} LR_{\text{CCI}} &= -2\ln\left(\frac{\pi^{n_{01}+n_{11}}(1-\pi)^{n_{00}+n_{10}}}{\pi_0^{n_{01}}(1-\pi_0)^{n_{00}}\pi_1^{n_{11}}(1-\pi_1)^{n_{10}}}\right) \\ &= -2\ln\left(\pi^{n_{01}+n_{11}}(1-\pi)^{n_{00}+n_{10}}\right) \\ &\quad + 2\ln\left(\pi_0^{n_{01}}(1-\pi_0)^{n_{00}}\right) + 2\ln\left(\pi_1^{n_{11}}(1-\pi_1)^{n_{10}}\right). \end{aligned} \tag{7.11}$$

For large N, LR_{CCI} follows a χ^2–distribution with one degree of freedom. The corresponding p–value is

$$p_{\text{CCI}} = 1 - F_{\chi_1^2}(LR_{\text{CCI}}). \tag{7.12}$$

The CCI test accepts the null hypothesis, H_0: the probability of failure at hour t is independent of the failure probability at hour $t-1$ (or, more concisely, H_0: $\pi_0 = \pi_1 = \pi$), if

$$p_{\text{CCI}} > p_{\text{test}}. \tag{7.13}$$

The significance level, $1 - p_{\text{test}}$, is typically set at 0.95.

The test statistic (7.11) is a sum of three log–likelihood terms, each of the form

$$L_{a,b} = \ln\left(p^{n_a}(1-p)^{n_b}\right), \tag{7.14}$$

where p, n_a, and n_b depend on the counts n_{ij}, $i = \{0,1\}$, $j = \{0,1\}$. For example, the second of the three terms in (7.11) is

$$L_{00,01} = \ln\left(\pi_0^{n_{01}}(1-\pi_0)^{n_{00}}\right) = \ln\left[\left(\frac{n_{01}}{n_{00}+n_{01}}\right)^{n_{01}}\left(\frac{n_{00}}{n_{00}+n_{01}}\right)^{n_{00}}\right]. \tag{7.15}$$

This term has the limits

$$\lim_{n_{00}\to 0} L_{00,01} = \lim_{n_{01}\to 0} L_{00,01} = 0.$$

These limits ensure that, if any of n_{00}, n_{01}, n_{10}, or n_{11} are zero, the corresponding log–likelihood terms in (7.11) are set to 0, keeping the test statistic LR_{CCI} well–defined.

Kupiec's frequency–based PoF test can be combined with Christoffersen's independence CCI test to form the conditional coverage (CC) test. The CC test statistic is

$$LR_{\text{CC}} = LR_{\text{PoF}} + LR_{\text{CCI}}. \tag{7.16}$$

For large N, LR_{CC} follows a χ^2–distribution with two degrees of freedom. The p–value is

$$p_{\text{CC}} = 1 - F_{\chi_2^2}(LR_{\text{CC}}), \tag{7.17}$$

where $F_{\chi_2^2}(\cdot)$ is the CDF of the χ^2–distribution with two degrees of freedom. The CC test accepts the null hypothesis, H_0: $\pi_0 = \pi_1 = \pi = \alpha$, if

$$p_{\text{CC}} > p_{\text{test}}. \tag{7.18}$$

7.1.5 Haas's Tests

Let $t_1 < t_2 < \ldots < t_{i-1} < t_i < \ldots < t_x$ denote the sequence of times at which VaR failures are seen in the sample of size N. Haas (2001) extended the TUFF test to test the time between successive failures $i-1$ and i (with the TUFF test used to test the time to first failure $i = 1$). This extension is referred to as the time between failures independence (TBFI) test. Let x represent the number of failures, and $n_i, i = 1, \ldots, x$ denote the number of hours between failure $i-1$ and i, where n_1 is the time to the first failure as defined in Section 7.1.3. Using the time between failures as a metric, the TBFI test evaluates whether failures are independent. A log–likelihood ratio based on (7.8) is defined for each n_i

$$\begin{aligned} LR_{\text{TBFI},i} &= -2\ln\left(\frac{\alpha(1-\alpha)^{n_i-1}}{(\frac{1}{n_i})(1-\frac{1}{n_i})^{n_i-1}}\right) \\ &= -2\left[\ln(n_i\alpha) + (n_i-1)\ln\left(\frac{n_i(1-\alpha)}{n_i-1}\right)\right]. \end{aligned} \tag{7.19}$$

Assuming that failures are independent, the overall log–likelihood ratio is the sum of the individual log–likelihood ratios,

$$LR_{\text{TBFI}} = \sum_{i=1}^{x} LR_{\text{TBFI,i}}, \tag{7.20}$$

which, for large N, follows a χ^2–distribution with x degrees of freedom. Consequently, the p–value of the test statistic is

$$p_{\text{TBFI}} = 1 - F_{\chi_x^2}(LR_{\text{TBFI}}), \tag{7.21}$$

where $F_{\chi^2_x}(\cdot)$ is the CDF of a χ^2–distribution with x degrees of freedom. The TBFI test accepts the null hypothesis, H_0: the times between successive failures are independent, if

$$p_{\text{TBFI}} > p_{\text{test}}. \tag{7.22}$$

If no failures are observed ($x = 0$), the scenarios discussed in the TUFF test apply.

Kupiec's PoF test can be combined with Haas's TBFI test to form the time between failures (TBF) mixed test, which has the test statistic

$$LR_{\text{TBF}} = LR_{\text{PoF}} + LR_{\text{TBFI}}, \tag{7.23}$$

and, for large N, follows a χ^2–distribution with $x + 1$ degrees of freedom. The p–value for this statistic is

$$p_{\text{TBF}} = 1 - F_{\chi^2_{x+1}}(LR_{\text{TBF}}). \tag{7.24}$$

The TBF test is accepted if

$$p_{\text{TBF}} > p_{\text{test}}. \tag{7.25}$$

If no failures are observed, the TBF test result is accepted only if the separate PoF and TBFI test results are also accepted.

The binomial, traffic light, PoF and TUFF tests are tests of unconditional coverage, while CCI and TBFI are tests of independence. The CC and TBF tests aim to evaluate both unconditional coverage and independence within a single test, and would appear to be more comprehensive than the single–characteristic tests. However, as we show in Section 7.2, the CC test can be accepted even though the PoF test (run alone) is accepted while the CCI test (run alone) is rejected. Similarly, the TBF test can be accepted even though PoF is accpeted while TBFI is rejected.

Suppose the CC test fails. Then either, or both, of the PoF and CCI components of the mixed test have failed, but knowledge of which requires running the PoF and CCI tests separately. Stated differently, if the CC test fails then the method being used to evaluate the VaR (which can include the active management practice of the portfolio) fails either, or both of, the conditional coverage or independence tenets outlined by Christoffersen (1998) – but the CC test fails to differential between them. The same issue holds for the TBF test.

A recent and more extensive review of VaR–based backtesting methods can be found in Zhang and Nadarajah (2018). They reviewed 28 tests, grouped into four categories: unconditional, conditional, independence and "other" approaches. They also identified software packages, both freeware and commercial, providing VaR backtesting capabilities.

7.2 Backtest Results

We apply the eight backtest methods to optimized portfolios from Chapters 4–6. For seven methods, excepting the TL test, we use $p_{\text{test}} = 0.05$ as the threshold for accepting or rejecting the test result.

7.2.1 Historical Optimization

For brevity, our analysis of the historically optimized portfolio discussed in Chapter 4 focuses on the TVP and T99 optimizations under a long–only strategy subject to a 4% turnover constraint. For comparison, we also perform backtesting on the unoptimized, benchmark portfolio EWBH.[3]

Fig. 7.1 plots the time series of hourly returns $r_p(t)$ and VaR values $\text{VaR}_c(r_p, t)$, $c =$ 95, 99, and 99.5, for these two optimizations and for the EWBH portfolio. The formal definition (3.23) follows the convention that VaR should be positive for losses (negative return values). In Fig. 7.1 and subsequent VaR plots, we drop that convention and report negative VaR values for losses. This allows for direct comparison of VaR values with the return time series plotted in the same figure. Consequently, for a fixed value of t, $\text{VaR}_{99.5}(r_p, t) < \text{VaR}_{99}(r_p, t) < \text{VaR}_{95}(r_p, t)$. For a given value c, VaR_c for EWBH had less variation and a smaller magnitude than for the optimized portfolios.

What is notable is that the difference $\text{VaR}_{95}(r_p, t) - \text{VaR}_{99}(r_p, t)$ is much larger than the difference $\text{VaR}_{99}(r_p, t) - \text{VaR}_{99.5}(r_p, t)$. This greatly affects the difference in the number of times (number of failures) that the return falls below $\text{VaR}_{95}(r_p, t)$ compared to the number of times the return falls below either $\text{VaR}_{99}(r_p, t)$ or $\text{VaR}_{99.5}(r_p, t)$. This is a key motivation for performing backtesting on $\text{VaR}_{99}(r_p, t)$ and $\text{VaR}_{99.5}(r_p, t)$ levels and the decision to change the Basel II accords' $\text{VaR}_{99}(r_p, t)$ backtest to the Basel III accords' $\text{VaR}_{99}(r_p, t)$ backtest.

Tab. 7.2 provides a summary of the failure statistics for these two optimizations and EWBH. The definitions of the variables corresponding to each column are detailed in Sections 7.1.1–7.1.3. With a moving window size of 2,000 hours, there were 1,000 moving windows in the data set, hence $N = 1,000$. The number of observed failures x dropped significantly from $1-\alpha = 0.95$ to 0.99; consequently the values n_{10} and n_{01} dropped commensurately, while n_{00} increased. Note that $n_{10} = n_{01}$ for each separate backtest. (This is also true of all the backtest results considered below.) The number n_{11} of consecutive failures was very small.

Tab. 7.3 summarizes the p–values calculated for all eight tests described in Sections 7.1.1–7.1.5 for these dynamic TVP and T99 optimizations and for the EWBH. (For the traffic light test, no p–value is computed, rather the significant value is the cumulative

3 We avoid considering the EWP benchmark as being unrealistic in terms of turnover cost.

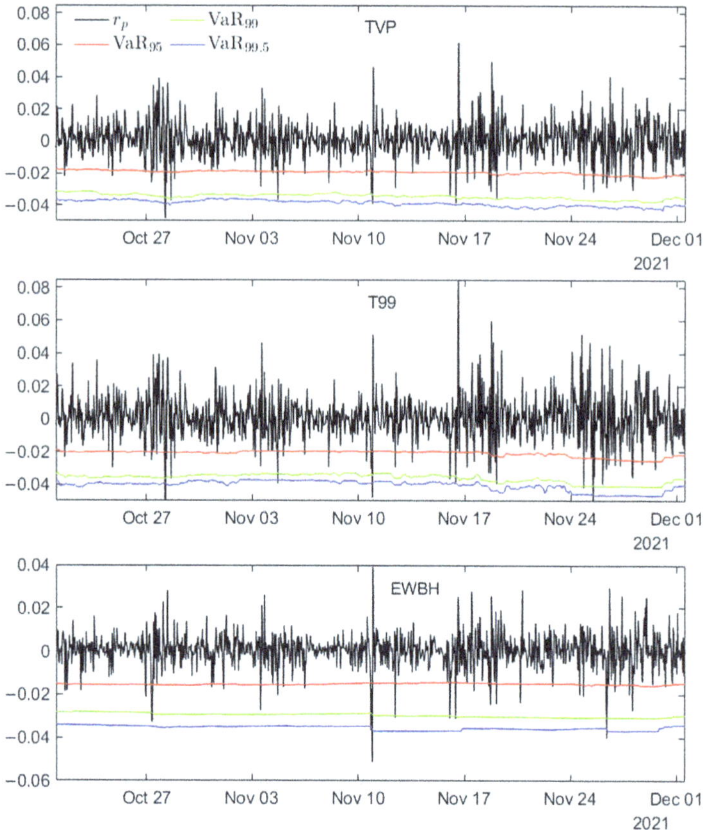

Fig. 7.1: Hourly returns $r_p(t)$ and VaR values $\mathrm{VaR}_c(r_p, t)$, $c = 95, 99$, and 99.5, for the historical TVP and T99 optimizations under the long-only strategy subject to a 4% turnover constraint, as well as for the benchmark EWBH portfolio

probability $P(X \leq x \mid N, \alpha)$.) To enhance visual clarity, we extend the red–yellow–green color scheme of the traffic light test to the other tests, with red indicating test rejection and green indicating acceptance.

For both optimizations, the unconditional coverage tests were accepted (traffic light test is green) for all three values of α. For the independence tests, CCI and TBFI, five of the six backtests were rejected for TVP while only three of six were rejected for T99. For the mixed tests, CC and TBF, four of six were rejected for TVP while only one was rejected for T99. Note that for the VaR_{99} test for TVP, the CC test was accepted in spite of the fact that the individual PoF test was accepted while the corresponding CCI test was rejected. The analogous problems occurred for T99 under the VaR_{99} TBF test and the $\mathrm{VaR}_{99.5}$ CC test.

In contrast, for the passive portfolio EWBH, the binary and PoF unconditional coverage tests were rejected at the VaR_{95} level. Similar to TVP, EWBH failed five of the six

Tab. 7.2: Failure statistics for the historical TVP and T99 optimizations under the long-only strategy subject to a 4% turnover constraint, as well as for the EWBH portfolio

$1-\alpha$	N	x	$1-\frac{x}{N}$	$N\alpha$	$\frac{x}{N\alpha}$	n_1	n_{00}	n_{10}	n_{01}	n_{11}
					TVP					
0.95	1000	47	0.953	50	0.94	15	908	44	44	3
0.99	1000	8	0.992	10	0.80	176	984	7	7	1
0.995	1000	3	0.997	5	0.60	176	994	2	2	1
					T99					
0.95	1000	45	0.955	50	0.9	15	911	43	43	2
0.99	1000	10	0.990	10	1.0	176	980	9	9	1
0.995	1000	6	0.994	5	1.2	176	988	5	5	1
					EWBH					
0.95	1000	35	0.965	50	0.7	15	934	30	30	5
0.99	1000	8	0.992	10	0.8	155	984	7	7	1
0.995	1000	3	0.997	5	0.6	503	993	3	3	0

backtests for CCI and TBFI and four of the six tests for CC and TBF. In contrast to TVP, EWBH had better passing rates for VaR_c as c increased from 95 to 99.5. Again, at VaR_{99} the mixed CC test for EWBH showed an acceptance even though the individual CCI test was rejected.

7.2.2 Dynamic Optimization

We now examine the backtest results for the dynamically optimized portfolios discussed in Chapter 5. Figure 7.2 presents the time series of hourly returns, $r_p(t)$ and the VaR projections, for the TVP and T99 optimizations under a long–only strategy subject to a 4% turnover constraint. The $\text{VaR}_\alpha(r_p, t)$ values were calculated using the sample of $S = 10,000$ simulated projections of $r_p(t)$ generated for each rolling window during the optimization.

Compared to the historical optimizations, the VaR values based on the larger set of simulated $r_p(t)$ values showed greater hourly sensitivity. The dynamic VaR results showed significant risk for the time periods around 28–29 October, 3–4 November, 10–11 November, 18–19 November and 24–25 November. The historical VaR results in Fig. 7.1 showed no significant increase in VaR for these dates. Note that the tail–optimized T99 predicted a much more significant VaR on 18–19 November than did the variance optimized TVP.

Tab. 7.4 provides a summary of the failure statistics for the VaR backtests. The trends are qualitatively the same as for the historical optimizations with two notable exceptions for the dynamic optimizations: the number of observed values x always equaled

Tab. 7.3: Summary of p-values for the backtests run on the historical TVP and T99 optimizations under the long-only strategy subject to a 4% turnover constraint, and on the EWBH portfolio

$1-\alpha$	p_{BIN}	$P(X \leq x \mid N, \alpha)$	p_{PoF}	p_{TUFF}
		TVP		
0.95	0.663	0.366	0.660	0.778
0.99	0.525	0.332	0.510	0.531
0.995	0.370	0.264	0.333	0.900
		T99		
0.95	0.468	0.261	0.461	0.778
0.99	1.000	0.583	1.000	0.531
0.995	0.654	0.763	0.664	0.900
		EWBH		
0.95	$2.95 \cdot 10^{-2}$	$1.42 \cdot 10^{-2}$	$2.17 \cdot 10^{-2}$	0.778
0.99	0.525	0.332	0.510	0.635
0.995	0.370	0.264	0.333	0.275

$1-\alpha$	p_{CCI}	p_{TBFI}	p_{CC}	p_{TBF}
		TVP		
0.95	0.596	$2.21 \cdot 10^{-2}$	0.789	$2.67 \cdot 10^{-2}$
0.99	$4.96 \cdot 10^{-2}$	$3.54 \cdot 10^{-2}$	0.117	$4.92 \cdot 10^{-2}$
0.995	$4.23 \cdot 10^{-3}$	$1.23 \cdot 10^{-2}$	$1.05 \cdot 10^{-2}$	$1.86 \cdot 10^{-2}$
		T99		
0.95	0.984	$9.69 \cdot 10^{-3}$	0.762	$1.12 \cdot 10^{-2}$
0.99	$8.47 \cdot 10^{-2}$	$4.30 \cdot 10^{-2}$	0.226	$6.49 \cdot 10^{-2}$
0.995	$2.46 \cdot 10^{-2}$	$9.28 \cdot 10^{-2}$	$7.29 \cdot 10^{-2}$	0.136
		EWBH		
0.95	$6.52 \cdot 10^{-3}$	$7.15 \cdot 10^{-4}$	$1.78 \cdot 10^{-3}$	$2.48 \cdot 10^{-4}$
0.99	$4.96 \cdot 10^{-2}$	$3.01 \cdot 10^{-3}$	0.117	$4.75 \cdot 10^{-3}$
0.995	0.893	$2.33 \cdot 10^{-2}$	0.620	$3.36 \cdot 10^{-2}$

or exceeded the expected number $N\alpha$; and the number of consecutive failures n_{11} was noticeable larger for $1-\alpha = 0.95$, and was zero in three out of the six backtests.

Tab. 7.5 summarizes the test results for these two dynamic optimizations. As in the historical optimizations, the unconditional coverage tests were all passed (although the TVP VaR_{99} traffic light test just fell into the yellow zone). The TVP optimization passed all eight tests for $1-\alpha = 0.995$, while the T99 optimization passed all tests for $1-\alpha = 0.99$ and 0.995.

Examining the test results for both historical and dynamic optimization leads to the conclusion that the unconditional coverage tests are easier to pass than the indepen-

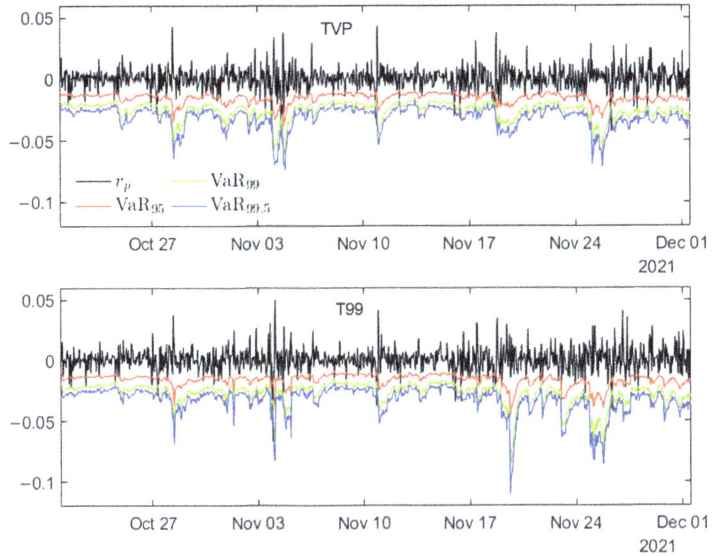

Fig. 7.2: Hourly returns $r_p(t)$ and VaR values $\mathrm{VaR}_c(r_p, t)$, $c = 95, 99$, and 99.5, for the dynamic TVP and T99 optimizations under the long-only strategy subject to a 4% turnover constraint

Tab. 7.4: Failure statistics for the dynamic TVP and T99 optimizations under the long-only strategy subject to a 4% turnover constraint

$1-\alpha$	N	x	$1-\frac{x}{N}$	$N\alpha$	$\frac{x}{N\alpha}$	n_1	n_{00}	n_{10}	n_{01}	n_{11}
					TVP					
0.95	1000	52	0.948	50	1.04	15	908	39	39	13
0.99	1000	15	0.985	10	1.50	15	971	13	13	2
0.995	1000	7	0.993	5	1.40	155	985	7	7	0
					T99					
0.95	1000	50	0.950	50	1.0	15	906	43	43	7
0.99	1000	11	0.989	10	1.1	87	977	11	11	0
0.995	1000	6	0.994	5	1.2	155	987	6	6	0

dence tests – a direct consequence of volatility clustering. These results also indicate that the tail-risk T99 optimization has better performance in adjusting to the volatility clustering than does the variance-based TVP optimization.

Tab. 7.5: Summary of p-values for the backtests run on the dynamic TVP and T99 optimizations under the long-only strategy subject to a 4% turnover constraint

$1-\alpha$	p_{BIN}	$P(X \le x \mid N, \alpha)$	p_{PoF}	p_{TUFF}
		TVP		
0.95	0.772	0.649	0.773	0.798
0.99	0.112	0.952	0.139	0.143
0.995	0.370	0.867	0.398	0.806
		T99		
0.95	1.000	0.520	1.000	0.778
0.99	0.751	0.417	0.754	0.891
0.995	0.654	0.384	0.664	0.806
$1-\alpha$	**p_{CCI}**	**p_{TBFI}**	**p_{CC}**	**p_{TBF}**
		TVP		
0.95	$5.84 \cdot 10^{-7}$	$2.10 \cdot 10^{-6}$	$3.64 \cdot 10^{-6}$	$3.08 \cdot 10^{-6}$
0.99	$1.76 \cdot 10^{-2}$	$3.88 \cdot 10^{-4}$	$2.00 \cdot 10^{-2}$	$3.17 \cdot 10^{-4}$
0.995	0.753	$8.32 \cdot 10^{-2}$	0.666	0.102
		T99		
0.95	$1.20 \cdot 10^{-2}$	$8.47 \cdot 10^{-3}$	$4.27 \cdot 10^{-2}$	$1.09 \cdot 10^{-2}$
0.99	0.621	0.250	0.842	0.313
0.995	0.788	0.579	0.877	0.670

7.2.3 Robust Optimization

We conclude with an examination of the backtest results for distributionally robust optimization. We consider the optimizations with $\beta = 0.3$ and $\epsilon = \{0, 0.00025, 0.0005\}$, as the $\beta = 0.3$, $\epsilon = 0$ results are similar to the historical TVP optimization.

Fig. 7.3 shows the hourly returns and VaR values for these three optimizations. The overall behavior of the three optimizations is reminiscent of the historical TVP and T99 optimizations with slightly more variability in VaR values. For the distributionally robust optimizations, $\mathrm{VaR}_{95} \in [-0.02, -0.024]$; $\mathrm{VaR}_{99} \in [-0.34, -0.43]$; and $\mathrm{VaR}_{99.5} \in [-0.37, -0.52]$. Interestingly, the $\epsilon = 0.00025$ optimization shows the greatest VaR variation.

The failure statistics are provided in Tab. 7.6. In all cases, the number of observed failures x falls (well) below the expected number $N\alpha$ with ratios $x/(N\alpha)$ in the range $[0.2, 0.8]$. For $\epsilon = 0.0005$, only a single failure was recorded for $\mathrm{VaR}_{99.5}$ occurring at hour 355 (16:00 on 4 November 2021). For $\epsilon = 0.00025$, two failures were recorded for $\mathrm{VaR}_{99.5}$ occurring at hours 177 and 355 (06:00 on 28 October and 16:00 on 4 November).

Tab. 7.7 provides the test results for these optimizations. There were failures of unconditional coverage tests for $\epsilon = 0.00025$ and 0.0005. However, all eight tests were

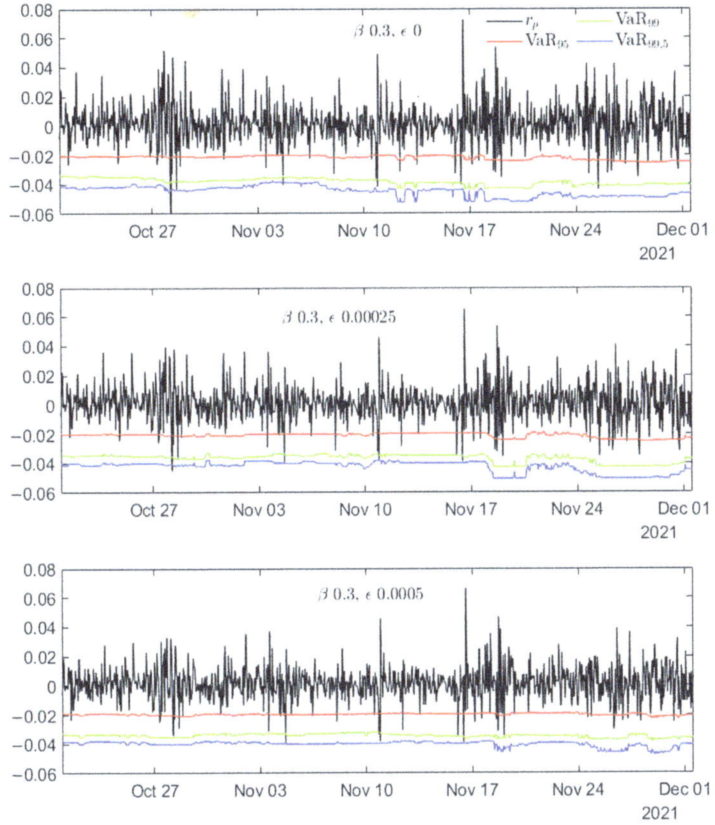

Fig. 7.3: Hourly returns $r_p(t)$ and VaR values $\mathrm{VaR}_c(r_p, t)$, $c = 95, 99$, and 99.5, for the distributionally robust optimizations with $\beta = 0.3$ and $\epsilon = 0, 0.00025, 0.0005$

passed at $\mathrm{VaR}_{99.5}$ for $\epsilon = 0.00025$ and at VaR_{99} for $\epsilon = 0.0005$. For $\epsilon = 0.0005$, at $\mathrm{VaR}_{99.5}$ the PoF test failed, yet both mixed tests CC and TBF passed.

Reviewing the backtests for the historical, dynamic and distributionally robust optimizations examined here, using as a criterion the requirement that all eight tests be passed, we have the following results:

dynamic, T99 – all passed for VaR_{99} and $\mathrm{VaR}_{99.5}$;
dynamic, TVP – all passed for $\mathrm{VaR}_{99.5}$;
distributionally, $\epsilon = 0.00025$ – all passed for $\mathrm{VaR}_{99.5}$; and
distributionally, $\epsilon = 0.0005$ – all passed for VaR_{99}.

Thus the backtest results on there own support dynamic, T99 optimization as a preferred portfolio management tool for crypto assets.

Tab. 7.6: Failure statistics for the distributionally robust optimizations with $\beta = 0.3$ and $\epsilon = \{0, 0.00025, 0.0005\}$

$1-\alpha$	N	x	$1-\frac{x}{N}$	$N\alpha$	$\frac{x}{N\alpha}$	n_1	n_{00}	n_{10}	n_{01}	n_{11}
					$\epsilon = 0$					
0.95	1000	40	0.960	50	0.8	15	923	36	36	4
0.99	1000	7	0.993	10	0.7	176	986	6	6	1
0.995	1000	4	0.996	5	0.8	176	992	3	3	1
					$\epsilon = 0.00025$					
0.95	1000	36	0.964	50	0.72	15	931	32	32	4
0.99	1000	6	0.994	10	0.60	176	988	5	5	1
0.995	1000	2	0.998	5	0.40	177	995	2	2	0
					$\epsilon = 0.0005$					
0.95	1000	38	0.962	50	0.76	15	927	34	34	4
0.99	1000	5	0.995	10	0.50	355	989	5	5	0
0.995	1000	1	0.999	5	0.20	355	997	1	1	0

Tab. 7.7: p-values for the backtests run on the distributionally robust optimized portfolio with $\beta = 0.3$ and $\epsilon = \{0, 0.00025, 0.0005\}$

$1-\alpha$	p_{BIN}	$P(X \leq x \mid N, \alpha)$	p_{PoF}	p_{TUFF}
		$\epsilon = 0$		
0.95	0.147	$8.06 \cdot 10^{-2}$	0.133	0.778
0.99	0.340	0.219	0.314	0.531
0.995	0.654	0.440	0.642	0.900
		$\epsilon = 0.00025$		
0.95	$4.22 \cdot 10^{-2}$	$2.11 \cdot 10^{-2}$	$3.29 \cdot 10^{-2}$	0.778
0.99	0.204	0.129	0.170	0.531
0.995	0.179	0.124	0.126	0.904
		$\epsilon = 0.0005$		
0.95	$8.17 \cdot 10^{-2}$	$4.33 \cdot 10^{-2}$	$6.95 \cdot 10^{-2}$	0.778
0.99	0.112	$6.61 \cdot 10^{-2}$	$7.86 \cdot 10^{-2}$	0.108
0.995	$7.29 \cdot 10^{-2}$	$4.01 \cdot 10^{-2}$	$2.85 \cdot 10^{-2}$	0.525

$1-\alpha$	p_{CCI}	p_{TBFI}	p_{CC}	p_{TBF}
		$\epsilon = 0$		
0.95	$9.21 \cdot 10^{-2}$	$5.07 \cdot 10^{-3}$	$7.84 \cdot 10^{-2}$	$4.06 \cdot 10^{-3}$
0.99	$3.59 \cdot 10^{-2}$	$4.94 \cdot 10^{-2}$	$6.66 \cdot 10^{-2}$	$5.69 \cdot 10^{-2}$
0.995	$8.95 \cdot 10^{-3}$	$4.62 \cdot 10^{-3}$	$2.95 \cdot 10^{-2}$	$9.33 \cdot 10^{-3}$
		$\epsilon = 0.00025$		
0.95	$4.44 \cdot 10^{-2}$	$2.36 \cdot 10^{-4}$	$1.36 \cdot 10^{-2}$	$9.75 \cdot 10^{-5}$
0.99	$2.46 \cdot 10^{-2}$	$3.84 \cdot 10^{-2}$	$3.12 \cdot 10^{-2}$	$3.36 \cdot 10^{-2}$
0.995	0.929	0.986	0.309	0.499
		$\epsilon = 0.0005$		
0.95	$6.49 \cdot 10^{-2}$	$6.80 \cdot 10^{-4}$	$3.51 \cdot 10^{-2}$	$4.09 \cdot 10^{-4}$
0.99	0.823	$9.14 \cdot 10^{-2}$	0.208	$5.03 \cdot 10^{-2}$
0.995	0.964	0.525	$9.08 \cdot 10^{-2}$	$7.42 \cdot 10^{-2}$

Bibliography

Andersen, T. G., Davis, R. A., Kreiß, J.-P., and Mikosch, T. V. (2009). *Handbook of Financial Time Series*. Springer-Verlag.

Artzner, P., Delbaen, F., Eber, J.-M., and Heath, D. (1999). Coherent measures of risk. *Mathematical Finance*, 9(3):203–228.

Baillie, R. T., Bollerslev, T., and Millelsen, H. O. (1996). Fractionally integrated generalized autoregressive conditional heteroskedasticity. *Journal of Econometrics*, 74(1):3–30.

Balkema, A. A. and De Haan, L. (1974). Residual life time at great age. *The Annals of Probability*, 2(5):792–804.

Ben-Tal, A., El Ghaoui, L., and Nemirovski, A. (2009). *Robust Optimization*. Princeton University Press.

Berlinger, E., Illés, F., Badics, M., Banai, Á., Daróczi, G., Dömötör, B., Gabler, G., Havran, D., Juhász, P., Margitai, I., et al. (2015). *Mastering R for Quantitative Finance*. Packt Publishing Ltd.

Bertsimas, D. and Sim, M. (2004). The price of robustness. *Operations Research*, 52(1):35–53.

Black, F. and Scholes, M. (1973). The pricing of options and corporate liabilities. *Journal of Political Economy*, 81(3):637–654.

Bloch, M., Guerard, J. B., Markowitz, H., Todd, P., and Xu, G. (1993). A comparison of some aspects of the US and Japanese equity markets. *Japan and the World Economy*, 5(1):3–26.

Bollerslev, T. (1986). Generalized autoregressive conditional heteroskedasticity. *Journal of Econometrics*, 31(3):307–327.

Broadie, M. and Glasserman, P. (1996). Estimating security price derivatives using simulation. *Management Science*, 42(2):269–285.

Campbell, S. D. (2005). A review of backtesting and backtesting procedures. Technical report, Finance and Economics Discussion Series, Federal Reserve Board, Washington, D.C.

Chan, W.-S. (2006). Outliers in nonstationary time series. *Journal of Quantitative Economics*, 4:75–83.

Cheridito, P. and Kromer, E. (2013). Reward–risk ratios. *Journal of Investment Strategies*, 3(1):3–18.

Christoffersen, P. F. (1998). Evaluating interval forecasts. *International Economic Review*, pages 841–862.

Cogneau, P. and Hübner, G. (2009). The 101 ways to measure portfolio performance. Available at SSRN 1326076.

Cont, R. (2001). Empirical properties of asset returns: Stylized facts and statistical issues. *Quantitative Finance*, 1(2):223.

Cox, J. C., Ross, S. A., and Rubinstein, M. (1979). Option pricing: A simplified approach. *Journal of Financial Economics*, 7(3):229–263.

D'Agostino, R. B., Belanger, A., and D'Agostino Jr., R. B. (1990). A suggestion for using powerful and informative tests of normality. *The American Statistician*, 44(4):316–321.

De Haan, L. and Peng, L. (1998). Comparison of tail index estimators. *Statistica Neerlandica*, 52(1):60–70.

Delage, E. and Ye, Y. (2010). Distributionally robust optimization under moment uncertainty with application to data-driven problems. *Operations Research*, 58(3):595–612.

DeMiguel, V., Garlappi, L., Nogales, F. J., and Uppal, R. (2009). A generalized approach to portfolio optimization: Improving performance by constraining portfolio norms. *Management Science*, 55(5):798–812.

Deng, Y. and Du, W. (2009). The Kantorovich metric in computer science: A brief survey. *Electronic Notes in Theoretical Computer Science*, 253:73–82.

Dickey, D. A. and Fuller, W. A. (1979). Distribution of the estimators for autoregressive time series with a unit root. *Journal of the American Statistical Association*, 74(366a):427–431.

Embrechts, P., Mikosch, T., and Klüppelberg, C. (1997). *Modelling Extremal Events: for Insurance and Finance*. Springer-Verlag.

Engle, R. F. (1982). Autoregressive conditional heteroscedasticity with estimates of the variance of United Kingdom inflation. *Econometrica*, 50(4):987–1007.

https://doi.org/10.1515/9781501517136-008

Esfahani, P. M. and Kuhn, D. (2018). Data-driven distributionally robust optimization using the Wasserstein metric: Performance guarantees and tractable reformulations. *Mathematical Programming*, 171:115–166.

Fabozzi, F. J. and Markowitz, H. M. (2011). *The Theory and Practice of Investment Management: Asset Allocation, Valuation, Portfolio Construction, and Strategies*. John Wiley & Sons.

Fama, E. F. and French, K. R. (1993). Common risk factors in the returns on stocks and bonds. *Journal of Financial Economics*, 33(1):3–56.

Fama, E. F. and French, K. R. (2010). Luck versus skill in the cross-section of mutual fund returns. *The Journal of Finance*, 65(5):1915–1947.

Fama, E. F. and French, K. R. (2015). A five-factor asset pricing model. *Journal of Financial Economics*, 116(1):1–22.

Ghouse, G., Khan, S. A., Rehman, A. U., and Bhatti, M. I. (2021). ARDL as an elixir approach to cure for spurious regression in nonstationary time series. *Mathematics*, 9(22):2839.

Granger, C. W. J. and Joyeaux, R. (1980). An introduction to long-memory time series models and fractional differencing. *Journal of Time Series Analysis*, 1(1):15–29.

Guerard, J. B., Chettiappan, S., and Xu, G. (2010). Stock-selection modeling and data mining corrections: Long-only versus 130/30 models. In *Handbook of Portfolio Construction*, pages 621–648. Springer.

Haas, M. (2001). New methods in backtesting. Technical report, Financial Engineering Research Center, Center of Advanced European Studies and Research, Bonn.

Haeusler, E. and Segers, J. (2007). Assessing confidence intervals for the tail index by Edgeworth expansions for the Hill estimator. *Bernoulli*, 13(1):175–194.

Hamilton, J. D. (2020). *Time Series Analysis*. Princeton University Press.

Hardy, M. R. (2006). An introduction to risk measures for actuarial applications. Technical report, Construction and Evaluation of Actuarial Models Study Note, Casualty Actuarial Society and the Society of Actuaries.

Hasan, M. F. (2023). The buy and hold investment strategy: Building and evaluating passive portfolios from Iraq stock exchange. *Technium Business and Management*, 5:79–87.

Hill, B. M. (1975). A simple general approach to inference about the tail of a distribution. *The Annals of Statistics*, 3(5):1163–1174.

Hilpisch, Y. (2018). *Python for Finance: Mastering Data-Driven Finance*. O'Reilly.

Hosking, J. R. M. (1981). Fractional differencing. *Biometrika*, 68(1):165–176.

Hu, Y., Lindquist, W. B., Rachev, S. T., and Fabozzi, F. J. (2024). Option pricing using a skew random walk pricing tree. *Journal of Risk and Financial Management*, 17:138.

Hull, J. C. (2022). *Options, Futures, and Other Derivatives*. Pearson.

Inui, K. and Kijima, M. (2005). On the significance of expected shortfall as a coherent risk measure. *Journal of Banking & Finance*, 29(4):853–864.

Jacobs, B. I., Levy, K. N., and Starer, D. (1999). Long-short portfolio management: An integrated approach. *Journal of Portfolio Management*, 25:23–32.

Jarrow, R. and Rudd, A. (1983). *Option Pricing*. Dow Jones-Irwin.

Jensen, M. C. (1968). The performance of mutual funds in the period 1945-1964. *The Journal of Finance*, 23(2):389–416.

Jorion, P. (2007). *Value at Risk: The New Benchmark for Managing Financial Risk*. McGraw-Hill.

Kaabar, S. (2024). *Deep Learning for Finance: Creating Machine & Deep Learning Models for Trading in Python*. O'Reilly Media, Inc.

Kantorovich, L. V. (1939). Mathematical methods of organization and planning. *Management Science*, 6(4):366–422.

Kirchgässner, G., Wolters, J., and Hassler, U. (2012). *Introduction to Modern Time Series Analysis*. Springer-Verlag.

Krokhmal, P., Palmquist, J., and Uryasev, S. (2002). Portfolio optimization with conditional value-at-risk objective and constraints. *Journal of Risk*, 4:43–68.

Kupiec, P. (1995). Techniques forverifying the accuracy of risk management. *Journal of Derivatives*, 3(2).

Lauria, D., Lindquist, W. B., Rachev, S. T., and Hu, Y. (2023). A binary tree, dynamic asset pricing model to capture moving average and autoregressive behavior. Available at arXiv:2304.02356.

Lin, F., Fang, X., and Goa, Z. (2022). Distributionally robust optimization: A review on theory and applications. *Numerical Algebra, Control and Optimization*, 12(1):159–212.

Lindquist, W. B., Rachev, S. T., Hu, Y., and Shirvani, A. (2022). *Advanced REIT Portfolio Optimization: Innovative Tools for Risk Management*. Springer.

Lo, A. W. (1991). Long-term memory in stock market prices. *Econometrica*, 59(5):1279–1313.

Lo, A. W. and Patel, P. N. (2008). 130/30: The new long-only. *Journal of Portfolio Management*, 34(2):12.

Lux, T. (2009). Stochastic behavioral asset-pricing models and the stylized facts. In *Handbook of Financial Markets: Dynamics and Evolution*, pages 161–215. Elsevier.

Makarov, I. and Schoar, A. (2020). Trading and arbitrage in cryptocurrency markets. *Journal of Financial Economics*, 135(2):293–319.

Mandelbrot, B. B. and Hudson, R. L. (2004). *The (Mis)Behavior of Markets: A Fractal View of Risk, Ruin, and Reward*. Basic Books.

Manuca, R. and Savit, R. (1996). Stationarity and nonstationarity in time series analysis. *Physica D: Nonlinear Phenomena*, 99(2–3):134–161.

Markowitz, H. M. (1952). Portfolio selection. *Journal of Finance*, 7(1):71–91.

Mazaheri, M. (2008). Risk budgeting using expected shortfall (CVaR). Available at SSRN 1150929.

McNeil, A. J., Frey, R., and Embrechts, P. (2015). *Quantitative Risk Management: Concepts, Techniques and Tools – Revised Edition*. Princeton University Press.

Merton, R. C. (1973). Theory of rational option pricing. *The Bell Journal of Economics and Management Science*, 4(1):141–183.

Monge, G. (1781). Mémoire sur la théorie des déblais et des remblais. *Histoire de l'Académie Royale des Sciences de Paris, avec les Mémoires de Mathématique et de Physique pour la même année*, pages 666–704.

Pickands III, J. (1975). Statistical inference using extreme order statistics. *The Annals of Statistics*, 3(1):119–131.

Rachev, S. T. (1985). The Monge-Kantorovich mass transference problem and its stochastic applications. *Theory of Probability and its Applications*, 29(4):647–676.

Rachev, S. T., Asare Nyarko, N., Omotade, B., and Yegon, P. (2024). Bachelier's market model for ESG asset pricing. *Journal of Risk and Financial Management*, 17(12):553.

Rachev, S. T. and Mittnik, S. (2000). *Stable Paretian Models in Finance*. John Wiley & Sons.

Rachev, S. T. and Rüschendorf, L. (1998a). *Mass Transportation Problems, Volume I: Theory*. Springer.

Rachev, S. T. and Rüschendorf, L. (1998b). *Mass Transportation Problems, Volume II: Applications*. Springer.

Rachev, S. T., Stoyanov, S. V., and Fabozzi, F. J. (2008). *Advanced Stochastic Models, Risk Assessment, and Portfolio Optimization: The Ideal Risk, Uncertainty, and Performance Measures*. John Wiley & Sons.

Rockafellar, R. T. and Uryasev, S. (2000). Optimization of conditional value-at-risk. *Journal of Risk*, 2:21–42.

Rockafellar, R. T. and Uryasev, S. (2002). Conditional value-at-risk for general loss distributions. *Journal of Banking & Finance*, 26(7):1443–1471.

Rogers, L. C. G. and Zhang, L. (2011). An asset return model capturing stylized facts. *Mathematics and Financial Economics*, 5(2):101–119.

Sharpe, W. F. (1966). Mutual fund performance. *The Journal of Business*, 39(1):119–138.

Shimokawa, T., Suzuki, K., and Misawa, T. (2007). An agent-based approach to financial stylized facts. *Physica A: Statistical Mechanics and its Applications*, 379(1):207–225.

Soros, G. (2015). *The Alchemy of Finance*. John Wiley & Sons.

Sortino, F. A. and Satchell, S. (2001). *Managing Downside Risk in Financial Markets*. Butterworth-Heinemann.

Tanous, P. J. (1999). *Investment Gurus: A Road Map to Wealth from the World's Best Money Managers*. Penguin.

Tsay, R. S. (2010). *Analysis of Financial Time Series*. John Wiley & Sons.

Tütüncü, R. H., Toh, K.-C., and Todd, M. J. (2003). Solving semidefinite-quadratic-linear programs using SDPT3. *Mathematical Programming*, 95:189–217.

Vaseršteĭn, L. N. (1969). Markov processes over denumerable products of spaces, describing large systems of automata. *Problemy Peredači Informacii*, 5(3):64–72.

Wald, A. (1945). Statistical decision functions that minimize the maximum risk. *The Annals of Mathematics*, 46(2):265–280.

Yen, Y.-M. (2016). Sparse weighted-norm minimum variance portfolios. *Review of Finance*, 20(3):1259–1287.

Zhang, Y. and Nadarajah, S. (2018). A review of backtesting for value at risk. *Communications in Statistics - Theory and Methods*, 47(15):3616–3639.

www.ingramcontent.com/pod-product-compliance
Lightning Source LLC
Chambersburg PA
CBHW081517190326
41458CB00015B/5396
9781501520099